Securing Global Transportation Networks

A Total Security Management Approach

About the Authors

Luke Ritter is the CEO of Trident Global Partners and a specialist in commercial and military transportation operations and logistics. Following graduation from the U.S. Naval Academy and service in the USN, Mr. Ritter managed intermodal railroad and trucking operations for a container line, and provided transportation security technology solutions to government and commercial clients while working for a systems integrator. He is a part-time instructor at the U.S. Merchant Marine Academy's Global Maritime and Transportation School, and serves as a contributing scholar at the Heritage Foundation.

J. Michael Barrett is a terrorism and homeland security expert with an extensive background in military intelligence and national security. A former Fulbright Scholar to Ankara, Turkey, Mr. Barrett is currently a Fellow in Homeland Security at the Manhattan Institute and the President of Counterpoint Assessments, a terrorism preparedness and risk mitigation firm. He previously served as Senior Analyst for the War on Terrorism Branch of the Office of the Chairman of the Joint Chiefs of Staff and as Lead Intelligence Officer for the Special Operation/Combating Terrorism Office of the Assistant Secretary of Defense. He has published more than a dozen newspaper and journal articles and has been interviewed dozens of times on television and radio, including ABC, NBC, Fox News, FRONTLINE, New York Public Radio, Bill Reilly's The Radio Factor, and The Canadian Broadcasting Company.

Rosalyn Wilson is a Manager at Reality Based IT Services, Ltd. [RBIS], an information technology security firm and subsidiary of SYS Technologies. She is also an independent consultant with over 25 years of experience in the transportation and logistics industry focusing on identifying and analyzing key performance issues. Mrs. Wilson has extensive railroad industry experience, having served in various capacities for over 11 years at the Association of American Railroads [AAR]. She is the author of the *State of Logistics Report*, an annual benchmark for US logistics activity, and of *Transportation in America*, a compendium of transportation information.

Securing Global Transportation Networks

A Total Security Management Approach

Luke Ritter

J. Michael Barrett

Rosalyn Wilson

New York   Chicago   San Francisco   Lisbon
London   Madrid   Mexico City   Milan   New Delhi
San Juan   Seoul   Singapore   Sydney   Toronto

The **McGraw·Hill** Companies

Cataloging-in-Publication Data is on file with the Library of Congress.

6 7 8 9 10 QVR/QVR 1 9 8 7 6 5 4 3 2

ISBN-13:   978-0-07-147751-2
ISBN-10:   0-07-147751-9

*The sponsoring editor for this book was Larry Hager and the editorial supervisor was
Patty Mon. It was set in Fairfield Medium by International Typesetting and Composition.
The art director for the cover was Margaret Webster-Shapiro.*

Printed and bound by IBT Global.

McGraw-Hill books are available at special quantity discounts to use as premiums and
sales promotions, or for use in corporate training programs. For more information,
please write to the Director of Special Sales, Professional Publishing, McGraw-Hill,
Two Penn Plaza, New York, NY 10121-2298. Or contact your local bookstore.

To all of the stakeholders—private, public, academic, and others—that make up and support global transportation networks. It is our hope that your collective entrepreneurial spirit will find the best means of creating value through the application of Total Security Management. By doing this, you will validate investments in private sector efficiency and security, while ensuring the resiliency of the global value chain upon which so much of humanity has come to rely.

Each of the authors would also like to dedicate this book to our families whose patience and support were critical components of this and all of our various projects. Thanks for the love and support, Lori, Catherine, and Alvin.

Contents at a Glance

Contents

Foreword

In the days that followed the tragic events of the terrorist attacks of September 11, 2001, the way forward was clear. We could be afraid, or we could be ready. Today that same path is just as appropriate regarding the security challenges present within the complex network operations of modern transportation. More than ever, those challenges, coupled with the intersecting forces of globalization, require a new and innovative management mindset that will drive improved security, yet also add business value and thus incentive for those at work in today's multifaceted transportation industries.

Securing our global transportation networks is a daunting task. Although many dedicated professionals have proposed novel and promising transportation security solutions, a comprehensive framework that can be applied across all enterprises to treat security as a core business function has long remained conspicuously elusive—until now.

The development of *Total Security Management [TSM]* delivers a novel, powerful and insightful approach to transportation security. Its message and direction represent a new and critical call to action—a call that must be heeded if collectively we are to develop the appropriate business case for systematically addressing the vulnerabilities inherent in global trade and creating commercial value in the process. The TSM approach has the potential to improve the way that transportation firms, insurance professionals, market analysts, and even consumers evaluate individual firms and their security and resiliency practices.

Today's global transportation networks are characterized by significant mutual dependencies among intersecting critical infrastructures—railroads, ports, airports, communications links and other essential elements that comprise a vibrant and robust transportation system. This interconnectedness also creates vulnerabilities in global transportation networks. Consequently, the effects of disruptions—whether natural or man-made—can be and often are significantly more damaging on a global scale. In other words, the cascading effects of terrorist acts, natural disasters and other "discontinuous events" can be significantly magnified as a result of globalization.

This book offers a much-needed framework for mitigating such risks. It articulates what I believe could be a comprehensive solution for approaching security in the context of sound business practices. By defining security and value in their broadest interpretations, the authors posit an approach that supports a review of all relevant internal and external processes. Identifying areas where security overlaps with other business prerogatives—asset tracking, access control, inventory management, cargo screening, etc.—establishes a baseline from which internal and external stakeholders can evaluate, analyze and take action to ensure the long-term viability of a firm's security and resiliency.

The paradigm that now exists in transportation security is similar to the paradigm that existed when the now revered Dr. W. Edward Demming tried to convince the business world in the 1960s that quality mattered—and that Total Quality Management could be used to create value. The now well-accepted theory of Total Quality Management was not initially met with open arms in the United States. I suspect that there are many professionals in the transportation industry today who may not endorse security management as a core business function that can create value. I ask those individuals to read on.

The authors provide real-world applications of the TSM methodology and walk the reader through a discussion of what the future may hold for enterprises that embrace this concept—a future in which security is managed as an essential business value.

The solution for lasting security must be a collaborative enterprise. Any conversations and any form of dialogue that we can use to better understand each other's perspective and identify common solutions to common problems will enable the civilized world not only to be more security conscious but also more coordinated and appropriately prepared. Using TSM as a common framework to develop transportation security solutions, the global trade community has the potential to close some of the security gaps that exist and rapidly begin to realize a real return on investment from transportation security dollars.

It is time to seize that potential. It is time for the global trade community to redouble its efforts to be innovative where security is concerned, to work together and to focus on applying the fundamental principles and practices introduced in this book to the challenges that we all share. Malcolm McLean, the founding father of containerization—and as such arguably the individual who has had the greatest single impact on the transportation industry in the last 100 years—once said, "I don't have much nostalgia for anything that loses money." It is time to move beyond transportation security solutions that do not create value and instead embrace those that provide a real return on investment.

We all know that the stakes are high, but high stakes can also lead to exceptional returns. TSM can help realize such returns. My hope is that readers will embrace the concepts of the TSM approach offered within these pages and act upon them. For action will lead to the security and economic outcomes our citizens, employees, shareholders and our country both require and deserve.

—Tom Ridge
First U.S. Secretary of Homeland Security

Total Security Management

Total Security Management is the business practice of developing and implementing comprehensive risk management and security best practices for a firm's entire value chain. This includes an evaluation of suppliers, distribution channels, and internal policies and procedures in terms of preparedness for disruptive events such as terrorism, political upheaval, natural disasters, and accidents.

Has Your Organization Fully Implemented Total Security Management Practices?

Q: Does your firm manage security initiatives with the explicit goal of creating value for the organization?

Q: Has your company identified and assigned value creation metrics that correspond to your security initiatives in order to measure return on investment?

Q: Are measures to ensure appropriate security for protecting fixed assets, assets in transit, brand equity/goodwill, and human capital considered fundamental business practices for your organization?

Q: Are stakeholders throughout your enterprise value chain (internal and external) empowered to create value through security?

Q: Has your firm made an organizational and financial commitment to improving its security posture while creating value?

Q: Has your firm initiated reasonable actions to prepare for, avoid, and minimize discontinuous events?

Q: Is business continuity planning managed as a core function within your organization?

Q: Can your organization identify and justify a security-related value proposition to shareholders, stakeholders, regulatory agencies and business partners?

Answer: Even if the answer to all of these questions is 'yes,' you still have not fully completed implementation of Total Security Management, which is a process that requires continual improvement and constant reexamination of security business practices!

Introduction

The Global Stage

The economic impact of a significant prolonged disruption to the global transportation network, and the supply chains that it sustains, would be measured in the billions, if not trillions, of dollars. This is because of the inherent fragility of the system that delicately balances the scheduling and global movement of a staggering volume of goods each year, including the more than $1.4 trillion worth of goods and nine million cargo containers that enter the United States alone.[1] In fact, on any given day, transportation assets in the United States are used to move approximately eight million truckloads of freight across four million miles of highway; 1.5 million railcars traverse over 170,000 miles of track; 2,400 flights pass through about 400 airports; and roughly 325 seaports transfer more than 25,000 containers. And this is just for the United States; global figures are exponentially higher.

Ships, trucks, planes, and railcars, and the infrastructure they travel on, as well as the people that operate them and the army of brokers, intermediaries, forwarders, and operations professionals that coordinate the movement of freight, are all "mission critical" assets for the global transportation network. However, there is a growing concern that within this sea of moving parts lay critical security gaps and a lack of operational resiliency that could cause tremendous damage from any significant disruptive event, whether man-made or natural.

Taking a closer look at deliberate events, it is widely accepted by industry analysts and security experts that the global transportation network is at considerable risk of being involved in future terrorist attacks. In an industry survey a full 70 percent of transportation company executives rate shipping containers as vulnerable to security risks, which is a sobering thought, given that an estimated 90 percent of the world's cargo now moves by container.[2]

To put the actual economic risk of a dirty bomb being detonated in a container at a U.S. port facility in perspective, the consulting firm Booz Allen Hamilton hosted a war game for 85 government and private sector transportation practitioners in October of 2002. Following the simulated detonation of the dirty bomb the war game's decision makers recommended that the government shut down every port in the United States for eight days, and some ports for twelve days. Analysis later determined that it would take up to three months to clear the shipping backlog from such an event and, even with these relatively short port closures; the estimated cost to United States businesses alone would exceed $58 billion.[3] This figure includes spoilage, lost sales/contracts, and manufacturing slowdowns, but does not account for the longer-term cascading global costs or the toll on the United States or foreign governments. Indeed, Deloitte Research paints an even starker picture, estimating that the total cost to the entire global supply chain of a weapon of mass destruction (WMD) delivered by a shipping container could be nearly $1 trillion.[4]

Although portions of the global transportation network have been heavily scrutinized since the terrorist attacks of 9/11 and a host of government and regulatory measures have been enacted, the fact remains that nearly all experts and analysts agree that the global transportation system remains vulnerable to a significant terrorist event, with many fearing that such an event is likely to have a devastating and lasting effect upon the entire system of global trade.

One doesn't need to fixate on terrorism to conclude that there are significant risks facing today's global transportation system, however. Incredibly, at the very same time the Booz Allen Hamilton war game was being carried out, a labor dispute affecting all U.S. ports on the West Coast resulted in a nine day port closure that caused losses conservatively estimated at $2 billion per day, including stoppages at several automobile manufacturing plants. Similarly, a truckers' strike in Great Britain during September of 2000 included the blockade of fuel depots and refineries and created shortages at British fueling stations, which then precluded the delivery of perishable goods to stores and brought the British government to within four days of having to ration food.[5] Significantly, the economic impact of such events was lower than what could be expected following a terrorist attack because there was no crime scene to investigate, no contamination to deal with, and no concern over follow-on events.

Climactic events can also cause major disruptions, such as in August 2005 when Hurricane Katrina closed U.S. ports in the Gulf Coast region and destroyed or disabled other critical transportation infrastructure in Texas, Louisiana, Alabama, and Mississippi. As noted by David M. Katz of CFO.com, "Even for companies far from the storm, Katrina may eventually result in billions of dollars in lost revenue...lost profits resulting from broken supply chains and absent business customers, as well as from the lengthy disruption of one of the country's largest commercial ports."[6] Indeed, with the ports and the transportation routes they serve closed for several weeks and degraded for many months, post-Katrina analysis showed total waterborne exports to the region fell $1.2 billion during September versus the previous month, led by losses in New Orleans ($649 million), Houston ($540 million) and Gulfport ($126 million), while crude oil imports to the storm-hit region during September 2005 were down 26 percent compared with September 2004.[7]

Tremendous costs and damage to transportation infrastructure also accompanied the Southern California wildfires in October and November of 2003, when more than 743,000 acres of brush and timber burned, over 3,700 homes were destroyed, and over $2.5 billion damage was caused. Similarly, the Northeast U.S. power outages of August 14th–15th 2003 offer yet another example of the impact of unforeseen events with huge consequences. The outages were caused by the simple lack of proper maintenance of trees alongside power lines, but forced the closure of several manufacturing plants because their just-in-time supplies of parts had been disrupted. Total economic impact from the outages and associated lost business has been estimated at $118 billion.[8] Even more worrisome on the natural disaster front is the fact that many public health experts now fear the mutation of the naturally occurring H5N1 "avian" flu virus into a strain that can be transmitted among humans, potentially resulting in another global influenza pandemic like the one in 1918, which killed more than 20 million people.

As illustrated by the Booz Allen Hamilton war game described previously, some of the most pressing challenges to conducting efficient global trade might come not from the event itself, but from the response of the affected governments. Whether the trigger event is man-made, such as the use of a dirty bomb in a port or other transportation node, or a naturally-occurring event such as a global pandemic of avian flu, the evidence from 9/11 indicates that the United States Government's first response will likely be to try and halt means of further attack or cross-contamination by shutting down the affected transportation links. It stands to reason that other governments around the world could choose to do the same, further increasing the impact of the disruptive event. The key players in the global transportation network need to develop a way to get ahead of this problem and to circumvent the threat by instituting new and improved security practices. If they do so today they can ensure that such practices are undertaken in the context of solutions that are complimentary to good business, however, with the goal of creating value for the firm kept as a primary objective. The alternative is to face the risk of additional regulatory actions that are unlikely to maintain this emphasis on business prerogatives or pragmatic concerns of efficiency.

The Challenge

Are terrorism, political upheaval, natural disasters, accidents, and other large-scale disruptive events happening more frequently, or are they just having a more significant impact than in previous years? It is our belief that the answer is yes—more disruptions are occurring and these disruptions are having a more significant impact. There has been a significant rise in the number of international terrorist incidents over the past decade, and at the same time severe weather that strikes anywhere on the globe is now more likely to threaten far-flung

Terrorism, political upheaval, natural disasters, accidents and other large-scale disruptive events are happening more frequently, and they are having a more significant impact than in previous years.

global supply chains, meaning that more incidents of more significance are occurring. The primary reasons are three-fold:

1. The impact of *globalization* and the attendant competitive forces of the global free market, which dictate that business processes be carried out by the lowest-priced provider wherever they may be located. This in turn increases the complexity and diversity of a firm's value chain.
2. The dramatic and growing *interconnectedness* and mutual dependencies *of global critical infrastructures* such as ports, highways, railroads, airports, telecommunications links, and power plants, coupled with the advent of lean business processes that minimize standing inventories, and in turn create increased collective risk from what would once have been relatively minor disruptions.
3. The continual threat of *discontinuous events* posed to this global and interconnected world by events that can severely disrupt normal patterns and cause changes in the free flow of goods including, but not limited to, severe weather, political upheaval, labor disputes, and terrorist attacks.

Today adverse events actually are more likely to happen as an outgrowth of the interplay among these three factors (which we refer to as the *three change agents* and define further in Chapter 1), and are more likely to have wider-ranging cascading effects as well. This phenomenon is the result of what noted enterprise vulnerability and business resiliency expert Yossi Sheffi calls *the high frequency of rare events,* explaining that, "while the likelihood for any one event that would have an impact on any one facility or supplier is small, the collective chance that some part of the supply chain will face some type of disruption is high."[9] Furthermore, with the increased likelihood comes increased severity. With each piece of the global transportation network increasingly tied to every other part, the cascading impacts from adverse events can now extend further than ever before.

The Solution: Total Security Management

What is the solution for firms faced with such pervasive and potentially existential risk to their business operations? We believe the best approach is to work through a framework that manages security as a core business function and integrates security prerogatives across all the activities of the enterprise. Doing so creates opportunities to create value in new and significant ways, which includes everything from cost savings from improved business processes and reduced theft from better asset management to enhanced brand equity or improved preparedness for catastrophic loss. In this manner, TSM can be used to turn security from a net cost into a net benefit. We call this approach *Total Security Management (TSM)*, a broad-based framework focusing on developing and implementing comprehensive risk management and security 'best practices' for a firm's entire value chain. TSM works with and through key internal and external stakeholders to ensure the use of a comprehensive approach to securing fixed assets, assets in transit, brand equity/goodwill, and human capital. It also emphasizes business continuity planning and an evaluation of the firm's suppliers, distribution channels, facilities selection criteria, and the internal policies and procedures that support preparedness for disruptive events.

Key Definitions

Total Security Management: The business practice of developing and implementing comprehensive risk management and security best practices for a firm's entire value chain.

This includes an evaluation of suppliers, distribution channels, and internal policies and procedures in terms of preparedness for disruptive events such as terrorism, political upheaval, natural disasters, and accidents.

Value Chain: "A high-level model of how businesses receive raw materials as input, add value to the raw materials through various processes, and sell finished products to customers. Value-chain analysis looks at every step a business goes through, from raw materials to the eventual end-user. The goal is to deliver maximum value for the least possible total cost."[10]

The intellectual foundations of the TSM approach draw from the seminal work of W. Edwards Deming and the manufacturing process improvements he championed through "Total Quality Management" (TQM) which, like our approach to total security, was based on the enterprise-wide application of best practices and measurable value creation. In fact, the benefits to be realized from using a TQM-like approach as an enabler of better security already can be found in the practices of some of America's best companies. For example, in 2004 Wayne Gibson Sr., Home Depot's Vice President for Global Logistics, testified before Congress that his firm, which at that time directly imported from 268 vendors and sourced 80 percent of its products from five countries, could use its current quality procedures, including inspections of its key vendors, to improve value chain security. He also noted that Home Depot could supplement existing anti-theft practices with anti-tamper efforts to improve container security.[11]

The relevance of this framework to security has been noted by several important research centers, including IBM's Global Movement Management practice, who concluded:

"The first critical challenge is integrating security and resilience with commercial imperatives of the system, including speed, performance, efficiency, interconnectedness, and predictability. Any new security-related activity will be more readily adopted if it enhances the performance and efficiency of the system, and will be resisted if it degrades that performance. The key private sector stakeholders in the system are likely to fund security investments only if they deliver concrete benefits beyond the often intangible benefits of security and resilience. If a good security concept is flawed in execution and harms the stakeholders who are responsible for implementing it on a day-to-day basis, then it will not be used–a worse outcome than having no security at all. Ultimately, security and resilience need to become embedded into these broader system imperatives, creating a culture of 'Total Security Management,' in a manner similar to the drive for 'Total Quality Management' in manufacturing, or the growth of safety as a paramount engineering norm in the aviation system and in automotive design."[12]

Similar to the 1970s and 80s challenge for firms to rethink manufacturing processes in order to make quality an integral part of each of the firm's business activities, the security and business continuity challenges facing today's firms require fresh thinking about how to identify and implement security improvement solutions in a manner consistent with the core business imperative of creating value from all actions and activities. In other words, firms need to find ways to implement security-relevant practices and procedures that also create value, whether in the form of labor savings, such as when certain automated technologies replace manual processes, or enhanced brand recognition, such as through championing the adoption of new end-to-end value chain security practices. The bottom line is that in business *security matters* and it should be managed to *create value*.

Firms need to find ways to implement security-relevant practices and procedures that also create value. The bottom line is that in business security matters and it should be managed to create value.

The Scope of TSM

Total Security Management refers to much more than the traditional security concerns of point defenses, security guards, and network passwords. Though these tactical considerations remain important, TSM focuses on security in the broader strategic sense by integrating security and resiliency initiatives into the

decision-making process on everything from vendor and third-party
logistics selection to the location of
outsourced production lines and call
centers. It does so in recognition
of the impact that the firm's broad
panoply of partners, customers, and
others can play in overall preparedness for disruptive events.

TSM focuses on security in the broader strategic sense by integrating security and resiliency initiatives into the decision-making process on everything from vendor and third-party logistics selection to the location of outsourced production lines and call centers.

Specifically, a firm's stakeholders
can be defined as those interested
and committed parties that interact
with or influence an enterprise and its activities. Collective
benefit is enabled by implementing a common understanding
of the enterprise's goals and those of all its partners, ideally
with the objectives and interests overlapping as much as possible. If this development of common objectives can be achieved,
enhancing and perfecting communication among stakeholders
can increase value for an organization and those with whom it
does business.

The number of relevant stakeholders for today's firms is
growing quickly for several reasons:

1. Global outsourcing brings additional relevant participants
   into each firm's value chain.
2. Understanding of critical infrastructure interdependencies
   brings greater interaction between any given enterprise
   and the operators of critical infrastructure.
3. Legal definitions imposed on publicly traded corporations
   require greater understanding of the role of management
   and shareholders.

The breadth and diversity of interests found among this
universe of stakeholders raises the importance of recognizing
that every interaction with each of these parties represents a
potential opportunity to create value for the firm. For this reason and to remind the reader of the broad scope of the firm's
stakeholders in terms of affecting value, wherever appropriate

this text uses the term *value chain* as opposed to the more limited concept of a *supply chain*.

The TSM Focus on Value Creation

The TSM approach was created to offer an all-hazards solution with the potential for appreciable process and efficiency gains that could offset the corresponding investment in new and more secure business practices. This approach of using a business case to justify the implementation of TSM is particularly important, for it draws out the reality that, at present, there is a marketplace failure in terms of properly rewarding firms that implement security best practices and take a dedicated approach to business continuity and resiliency. This oversight includes key industry players that may recognize the inherent worthiness of security initiatives, but nonetheless lack the framework for a structured enterprise-wide approach to implementing total security initiatives in ways that create value for their organization. This challenge extends beyond the transportation firms themselves, for the lack of an appropriate analytical framework also eludes many of the customers, industry analysts, investors, and insurers whose collective market-driven decisions can either validate or discount corporate decisions about total security.

Fortunately, TSM offers a framework that supports careful evaluation of the relative worthiness of a firm's approach to security and resiliency. Firms that implement security initiatives which are quantifiably more effective and more comprehensive than those of their competitors should expect to be rewarded accordingly by the marketplace. If verifiable, such competitive advantages should include benefits such as the ability to more rapidly reconstitute its value chain following a disruptive event. At the same time, implementation of TSM may serve as a deterrent to those who might otherwise attempt to disrupt a firm's value chain. The critical business process of measuring return on investment for security initiatives can provide valuable data to industry analysts, institutional investors, and insurers—all of which influence the firm's market valuation.

By highlighting existing (although often unaccounted for) risks, TSM helps define a firm's risk profile more fully, in order to foster better-informed risk management decisions, which in turn helps protect the enterprise from surprise. As the saying goes, "knowledge is power," and when knowledge of risk increases so do the opportunities to improve the firm's position, both relative to the risk itself and to peer competitors.

As for covering the costs of some of the TSM process improvements, some industry studies report that the combined savings in paperwork, manpower and theft reduction will cover most if not all of the costs associated with the new systems.[13] "The investment required to create a more secure supply chain is easily justified when compared to the costs associated with experiencing longer, unpredictable lead-times or acute disruptions. Among other things, these costs come in the form of:

· Additional inventory
· Slowing or shutting down production lines
· Lost revenue due to stock-outs or missed promotions
· Longer cash-to-cash cycles
· Higher insurance rates
· Increased transportation costs (e.g., more expedited shipments)"[14]

In fact, Deloitte Research asserts that during a test of a new smart racking system involving sixty-five companies across three continents, firms were able to document up to $400 in savings per shipment.[15] Obviously, there are a great number of variables at work here and the likelihood is that some but not all firms would find this 100% offset to be the case. However, even if one presumes the theft reduction from the new measures only cover half or a third of the system's implementation, the remaining costs can be off-set by other benefits of the improved processes, such as improved catastrophic preparedness, better integration amongst value chain partners, more timely knowledge of where products are in the supply network, and more insightful risk calculations for strategic growth and outsourcing decisions.

TSM in Action

The specific implementation of Total Security Management will vary from firm to firm and may require examining not only first tier but also second and third tier partners for critical functions; moving from a country or region-specific focus to a global one; and expanding crisis management planning to include business continuity planning as well. The Total Security framework is bounded by Five Strategic Pillars that define its central tenets for all security processes, including that they must:

· Create value for the firm
· Involve all relevant value chain partners
· Institute continual improvement
· Help avoid, minimize or survive discontinuous events, and
· Support business continuity plans

These pillars are in turn supported by Four Operational Enablers, which cover important focus areas for improvement including implementation of industry best practices, increased situational awareness, reliance upon training and exercises, and outreach to all relevant parties. In practice, as in the layout of this book, the initial application of the *Strategic Pillars* and the *Operational Enablers* is achieved at the *tactical* level by using the risk management approach to assess operations and processes across four critical functional areas: protection of fixed assets, assets in transit, brand equity/goodwill, and human capital. Relative resiliency and preparedness can then be benchmarked against other firms and process improvements can be analyzed in terms of their relevance to and conformity with the *TSM Value Creation Model*.

Conclusion

The world is changing all around us and the combined forces of globalization, infrastructure interdependence, and discontinuous events are providing incentives to make global transportation networks more resilient. The tipping point with regards to the need for Total Security has come, but it has come only recently. The timing is right for firms to begin managing security in a concerted and holistic fashion that creates value for the firm and a return on investment from reasonable security practices. Firms can utilize the value creation methodology outlined in this book to make rational and calculated security decisions in addressing and mitigating these concerns, and can do so in a manner that addresses security as a core business function.

The first four chapters of this book present the theoretical framework and primary tools of the Total Security Management approach. Chapter 1 begins with a thorough examination of the three change agents and the foundations of Total Security Management. Chapter 2 then moves on to a detailed discussion of TSM's structured, holistic approach as defined by the five strategic pillars, the four operational enablers, and the TSM Value Creation Model. Chapter 3 explores the business case that underpins TSM processes and discusses how TSM implementation can create value for firms that adhere to its precepts. Chapter 4 begins with an analysis of how transportation firms can apply TSM, including an explanation of the risk management process and how tools such as risk mapping help determine a firm's most appropriate risk mitigation strategy.

Chapters 5, 6, 7 and 8 are arranged so as to demonstrate how to apply the precepts of TSM to the specific asset categories of most concern to transportation firms, namely: fixed assets, assets in transit, brand equity/goodwill, and human capital. Chapter 9 addresses the critical process of Business Continuity Planning (BCP), which involves planning and preparedness to ensure the continuity of critical operational functions in the aftermath of a discontinuous event. Finally, the book concludes with Chapter 10, which offers analyses of the current and future status of TSM and taking a closer look at some of the ways in which the global marketplace is already taking the first steps in implementing bedrock TSM principles.

Notes

1. U.S. Customs and Border Protection, U.S. Department of Homeland Security, "Press Kit–An Introduction to U.S. Customs and Border Protection: Trade," 15 December 2005, <http://www.cbp.gov/linkhandler/cgov/newsroom/fact_sheets/press_kit/trade_press.ctt/trade_press.pdf> (21 April 2006).
2. Deloitte Research, *Prospering in a Secure Economy*, 2004, <http://www.deloitte.com/dtt/cda/doc/content/DTT_DR_ProsSecFull_Sept2004.pdf#search='Deloitte%20Research%20prospering%20in%20the%20secure%20economy> (22 April 2006). *See also* U.S. Customs and Border Protection, U.S. Department of Homeland Security, "Press Kit–An Introduction to U.S. Customs and Border Protection: Trade," 15 December 2005, <http://www.cbp.gov/linkhandler/cgov/newsroom/fact_sheets/press_kit/trade_press.ctt/trade_press.pdf> (22 April 2006).
3. Booz, Allen, Hamilton, *Threats to Port Security Call for Integrated Public/Private Action*, 4 December 2004, <http://www.boozallen.com/home/publications/article/1440496?lpid=661123> (April 29, 2006).
4. Deloitte Research, *Prospering in a Secure Economy*, 2004, <http://www.deloitte.com/dtt/cda/doc/content/DTT_DR_ProsSecFull_Sept2004.pdf#search='Deloitte%20Research%20prospering%20in%20the%20secure%20economy> (11 March 2006).
5. Yossi Sheffi, *The Resilient Enterprise* (Massachusetts: The MIT Press, 2005), 166.
6. David K. Katz, *Supply Chains in Katrina's Wake*, CFO.com, September 22, 2005, <http://www.cfo.com/article.cfm/4421125?f=related> (April 20, 2006).
7. John Peige, "Gulf Coast Huriccanes Have Huge Impact on Shipping Flows," *MM&P Wheelhouse Weekly*, 9 no. 43, (27 October 2005), <http://www.bridgedeck.org/mmp_news_archive/2005/mmp_news051027.html#anchor838850> (29 April 2006).
8. Ram Reddy, "Awake in the Dark, "*Intelligent Enterprise on the Web*, 1 January 2004, <http://www.intelligententerprise.com/040101/701infosc1_1.jhtml> (29 April 2004).
9. Yossi Sheffi, *The Resilient Enterprise* (Massachusetts: The MIT Press, 2005), 26.
10. *Investopedia.com*, "Dictionary: Term of the Day," (2006) <http://www.investopedia.com/terms/i/incentivetrust.asp> (29 April 2006).
11. Andrew K. Reese, "Building the Secure Supply Chain," *Supply & Demand Chain Executives on the Web*, (2005), <http://www.sdcexec.com/article_arch.asp?article_id=5287> (29 April 2006).
12. W. Scott Gould and Christian Beckner, IBM Business Consulting Services, Global Movement Management: Securing the Global Economy, November 2005, page 5.

13. Deloitte Research, *Prospering in a Secure Economy*, 2004, <http://www .deloitte.com/dtt/cda/doc/content/DTT_DR_ProsSecFull_Sept2004 .pdf#search='Deloitte%20Research%20prospering%20in%20the% 20secure%20economy> (21 April 2006). *See also* Hau L. Lee and Seungjin Whang, *Higher Supply Chain Security with Lower Cost: Lessons from Total Quality Management*, October 2003, <https:// gsbapps.stanford.edu/researchpapers/detail1.asp?Document_ID=2273> (21 April 2006).

14. Adrian Gonzalez, "Trade Security: A Wildcard in Supply Chain Management," ARC Advisory Group, September 2002, <http://72.14.203.104/ search?q=cache:czZiXTOgr_IJ:130.94.251.134/pdf/arc_security_ sept2002.pdf+ARC+Advisory+Group+September+2002+ Adrian+Gonzalez&hl=en&gl=us&ct=clnk&cd=9> (22 April 2006).

15. Deloitte Research, *Prospering in a Secure Economy*, 2004, <http://www .deloitte.com/dtt/cda/doc/content/DTT_DR_ProsSecFull_Sept2004 .pdf#search='Deloitte%20Research%20prospering%20in%20the% 20secure%20economy> (11 March 2006).

"Globalization: The inexorable integration of markets, nation-states and technologies to a degree never witnessed before–in a way that is enabling individuals, corporations and nation-states to reach around the world farther, faster, deeper, cheaper than ever before, and in a way that is enabling the world to reach into individuals, corporations, and nation-states farther, faster, deeper, cheaper than ever before."

—Thomas Friedman

Chapter One

Global Trade and Total Security Management

The world today is a truly astonishing place, one that people born just a few generations ago would scarcely recognize. Today, more people from more places can connect and interact with each other more easily than at any time in history. The advent of modern global transportation networks, satellite communications, and affordable personal computers all working together has allowed capitalism and economic competition to flourish amongst an ever-wider number of players, each simultaneously acting upon and reacting to the others. The effects of this global economic interaction, and the technological advances and far reaching changes that both feed into and grow out of the process, are fostering a corresponding revolution in business practices.

It has become commonplace to observe that the world is changing, that technology has altered our lives, and that we are in the midst of a revolutionary re-definition of how each part of the global economy impacts and interacts with all other parts. However, the global business community is still recognizing and comprehending the systemic economic and social effects of this redefinition process. One of the most significant results has to do with the realization that the complex and ever-expanding international "value chains" of all manner of firms have created a certain inherent fragility.

3

Value Chain

The totality of activities that add value to the business, including primary partners in the supply chain and entities that affect sales, marketing, infrastructure management, and other support functions.

These connections in the global economy can explain, in part, why a hurricane that slams into New Orleans can affect worldwide oil prices as well as America's foreign policy towards the government of Venezuela, why communications outages following monsoons in India affect the global processing of bank payments in New York and London, and why purported security issues can become decisive factors in a public uproar over a multi-billion dollar international maritime terminal operating deal.[1] Consequently, it is incumbent upon individuals and organizations to examine globalization's effects so that their responses can be correctly calibrated.

Currently, business practices that are put in place to help reduce the risks from this fragility are typically neither recognized nor measured by the consumers or analysts who specialize in evaluating a firm's relative value. This is problematic because, in order to responsibly manage the risks inherent in today's global economy, investments in the very resiliencies that mitigate against failures in the system need to be defined and measured. Once these measures are assigned an appropriate value, the marketplace will be empowered to reward those firms that institute best practices and to discount those that do not.

This, in essence, is the philosophical underpinning of *Total Security Management* (TSM), a novel approach to preparing for and managing the new security and business continuity risks created by globalization. In other words, TSM is about much more than just security—it's a unifying theory about the resiliency and survivability of the enterprise. As such, it redefines security as a key aspect of the firm's lasting value proposition taking security goals from the periphery of a firm's processes to the very core of the imperative to remain competitive in the face of change.

It is no mere coincidence that TSM is emerging now, at this time in history. In today's networked economy, companies simply must concern themselves with the security practices of their partners. According to Deloitte Research, "It typically takes twenty five different parties and thirty different documents to get goods from one end of the supply chain to the other. With all these handoffs, the opportunities for tampering are plentiful."[2] As a result, the potential benefits of implementing effective security practices have never been greater, but only because the threat of cascading costs due to a systemic interruption have never been greater. From earthquakes and hurricanes to labor strikes, civil wars, and terrorist attacks, governments around the world have proven time and time again that they cannot protect against, nor rapidly recover from, disruptive events everywhere and at all times. This means that the private sector must take more responsibility for preparing itself for the aftermath of future disruptive events. It is critical for the private sector, and the financial and insurance analysts that help shape its priorities, to actively address these risks by rewarding the preventive measures and efforts of those firms that more appropriately prepare for the realities of the world in which we live.

Similarly, a 2003 report from Great Britain's Cranfield School of Management declares,

> The world is alerted to risk and disaster in a way that would have seemed pessimistic and morbid just ten years ago. The combination of 'global' shocks that have been experienced in the last few years... has been a 'wakeup' call to the intrinsic vulnerability of our complex networked economies. In the context of the supply chain, the experience of disruption of supply from the variety of shocks has been felt in terms of social welfare, employment, economic activity, and ultimately in corporate and global wealth.[3]

The same report concludes, "Every enterprise is a complex network of suppliers and suppliers' suppliers that the company connects to its customers and then its customers' customers. The company may be connected not only to this specific supply chain community but also to a number of other supply chain networks. Furthermore, the physical logistics operations that connect the nodes in these chains are an integral part of the network—be it by road, rail, air or ship."

Analysts and academics that study these matters have begun to take note, fortunately, and often do so through research funded by major international firms. For example, David Closs and Edmund McGarrell, in a report for the IBM Center for the Business of Government, state:

> International and domestic incidents over the last three years have emphasized the need for an integrated approach to supply chain security management. Just as a chain is no stronger than its weakest link, a supply chain is only as secure as its weakest link, which includes the suppliers, manufacturers, wholesalers, retailers, carriers, terminals, and governmental institutions that plan, manage, facilitate, and monitor the global movement of goods.[4]

Similarly, in a study conducted by Stanford University's Global Supply Chain Management Forum titled: "Innovators in Supply Chain Security: Better Security Drives Business Value," companies were able to quantify business benefits associated with security initiatives—on average, these companies:

· Reduced Customs inspections by 48%
· Saw 29% reduction in transit times
· Improved asset visibility by 50%
· Improved on-time shipping to customers by 30%
· Reduced time taken to identify problems by 21%
· Reduced theft in inventory management by 38%
· Reduced excess inventory by 14%
· Reduced customer attrition by 26%.[5]

The first step that the global business community can take to manage this newly recognized and emerging risk is to examine its root causes. The ongoing processes that are leading to the redefinition of business practices and a realignment of global economic power have created both the need for and ensured the value of the *Total Security Management* approach. The three distinct but related **Change Agents** that embody the most significant aspects of the processes we are witnessing are:

1. Globalization
2. Infrastructure and Economic Interdependencies
3. Discontinuous Events

CHANGE AGENT #1: Globalization

Many of the tangible benefits of globalization are undeniable. Though widespread availability and use of laptop computers, e-mail, and wireless internet connections were barely imaginable just twenty-five years ago, today such tools are essential to the conduct of business. Indeed, in today's global economy, cellular telephones and pocket PCs can be designed in Japan to use parts manufactured in China, South Korea, and a dozen other countries, which are then shipped halfway across the world, assembled in Mexico, and sold in the United States. This creates economic integration on a truly astonishing level. One result of using such massive economies of scale and global outsourcing to find the lowest production costs and best value

Globalization

Economists, commentators, and notable scholars can and do disagree about the extent, causes, and even morality of this process, this 'globalization' of our world. Indeed, the very definition of globalization, and the extent of its influence, continue to change almost as fast as one can place a label on it. But no one denies that globalization is occurring. It appears that globalization, whether as defined by journalist and leading advocate Thomas Friedman at the opening of this chapter or by almost any other significant variant, is real, and is here to stay.

(wherever it may be) is that an increased number and diversity of goods are more broadly available and ensure a better standard of living for nearly every nation, comparatively, than at any time in the history of the world.

At the same time, jobs once thought to be safe have been exported overseas. The telecommunications revolution has connected the world with high-speed phone lines and digital connections that enable data processing from anywhere in the world at minimal cost. The global telecommunications infrastructure, and the internet that it sustains, have connected factory workers in China with end-users in Europe, and, increasingly, the lower wage but highly-skilled workforce of developing nations like India with global firms that are outsourcing. Many of these firms are now outsourcing nearly every conceivable piece of the production cycle that can be performed elsewhere, including the writing of computer code, answering customer service telephone calls, and even preparing tax returns for people in other countries.[6] Amazingly, according to a study by McKinsey & Company on the future of business process outsourcing, the current worldwide offshore market for such services exceeds $11 billion per year, but the potential market is between $120 billion and $150 billion.[7]

The reason such changes can occur is that the relentless competitive forces of the global free market dictate that everything must be done as efficiently as possible, for if a process can be performed cheaper somewhere else, and the current leading firms do not shift their practices accordingly, then they'll find that their competitors will. When a firm has the entire world to work with when selling its wares, new and massive efficiencies are found continually, as someone somewhere attempts to compete with the status quo by trying something new. With the increasingly more efficient global transportation system and nearly ubiquitous communications networks, just-in-time processing has in some cases become the only way to survive.

Because globalization plays such a fundamental role in creating the need for a Total Security Management approach, it is only fitting to take a moment to review its origins and explore its implications. According to the "Globalisation Guide," a

non-profit UK-based think tank that studies the effects of globalization, the first definable wave dates back to the sixteenth
century when European capitalists first circumnavigated the
globe and granted exclusive import licenses to private firms
such as the British East India Company. The process then
continued, slowly by today's standards but across undeniably
vast distances given the technologies of the time. In the late
1800s, with communications technologies like the transatlan-
tic telegraph cable in place and with international standards
developed for size, weight, and other measurements, there was
a major expansion in world trade and investment.[8]

The race for resources and global competition quickly fed
the flames that became World War I, and following the war's
end many governments retreated from open markets, preferring
to shelter themselves instead by enacting anti-trade protection-
ist policies, which in turn helped foster the Great Depression.
This period of economic integration was further hampered by
massive mobilizations during World War II. Nonetheless, the
Allied victory over the Axis Powers brought about another period
of rapid integration of the global economy and an alignment
heavily based on trade and cooperative development, at least
for the nations outside the Soviet Union's sphere of influence.
It was during this period that the rise of great multinational
companies was seen—companies that began to produce
similar or identical goods and services that they then sold
to consumers in many different nations. When added to the
incredible improvements in air travel and truly robust and reliable international communications, the world rapidly began to
shrink. Next, with the fall of the Berlin Wall in 1989 and the
subsequent collapse of the Soviet Union, there was a growing
perception of a triumph of capitalism over communism. The
stage was now set for even more rapid and complete economic
integration.[9] Finally, as noted globalization guru Thomas
Friedman observes, beginning in the 1990s and continuing
into the twenty-first century, the development of the internet
and a massive global investment in high speed data lines connected workforces in developing countries with firms based in
increasingly expensive developed nations, creating a marriage

made in heaven—at least in terms of efficiency and wholesale outsourcing of services.[10]

There is another side to the story, however, as some of the most beneficial aspects of globalization have brought unintended consequences. The quest for spectacular efficiency has inadvertently created a weakening of the system's capacity to absorb natural and man-made anomalies. Thus, despite tremendous cost savings for day-to-day business, current market forces have made the system even more interconnected, and have increased the potential systemic impacts of the failure of any single piece. The imperatives of daily business survival amid the forces of relentless competition have driven firms towards increasingly lean production, sales, and distribution cycles. These trends have in turn fostered business process reengineering to the point of minimum waste—all in order to remain competitive during periods of normalcy. This has had the unintended consequence of placing every other part of the system at risk by increasing the likelihood that a catastrophic event could trigger cascading systems failure. In the face of forces such as political instability, severe weather, growing oil dependency, terrorism, and looming confrontations over raw resources between developed and developing countries, the potential for a catastrophic disruption in business has become all too real.

Consider the plight of manufacturers operating in the just-in-time environment. These firms cannot remain competitive if they keep large quantities of parts inventory on hand because it is grossly inefficient to first receive, catalog, move, and store inventory, and then later locate, transport, and use these parts to assemble end products. It is much less expensive to have a system of parts shipments that arrive almost

> The quest for spectacular efficiency has inadvertently created a weakening of the system's capacity to absorb natural and man-made anomalies. This has had the unintended consequence of placing every other part of the system in ever-greater danger by increasing the risk of catastrophic cascading systems failure.

literally at the last second, virtually eliminating that middle step by coming off the delivery truck on one end of the factory, then straight onto the production line, and emerging as a fully assembled product. But whenever there is a massive disruption of their deliveries, these same firms tend to suffer more dramatic consequences. A recent transportation industry event that illustrated how sensitive production can be to certain dependent variables was the 2002 management lockout of U.S. dockworkers up and down the West Coast of the United States. This single labor dispute lasted for nine full days, upset the entire cargo handling system for twenty-nine ports, and consequently prevented the movement of a myriad of commodities, including auto parts. Lack of parts then forced the temporary closure of several U.S.-based automobile manufacturing plants. The port lockout cost the U.S. economy an estimated two billion dollars per day because the affected ports handle approximately twenty-one percent of all goods imported into the United States and about nine percent of all goods exported. The strike also forced some manufacturers who rely on foreign parts, including Boeing, GM, and Toyota, to shut down or realign their assembly processes because they lacked the resiliency to respond on short notice. In addition, produce was lost at sea as it spoiled while over 150 ships waited at anchor for opportunities to unload. After the strike was resolved it took weeks before the domestic maritime, rail, truck, and air transportation networks worked through the backlog. An important thing to note is that this was a disruption without destruction, a period of inactivity but without any of the damage to the infrastructure that would accompany severe storms, acts of terrorism, or the ravages of war.

Another major concern about globalization is the powerful but often unwelcome effect it can have upon traditional societies. Although economic development usually means improved healthcare and better quality of life, it often also means a changing of the priorities within a society. For example, in the traditionalist societies of the Middle East, massive social upheaval happens as nomadic peoples are forced to settle in one place. However, this upheaval also happens when conservative, deeply religious cultures feel imposed upon by dominant external factors that threaten to change their lives by challenging their cultural values. Because this isn't driven by any master plan but rather by the same economic forces that drive standardization across borders and communities (globalization, in other words), it often happens without regard for the people who would like to continue on traditional paths. (Indeed, some of the tensions that grow out of these processes are at the heart of current threats from terrorism, as discussed in Change Agent #3 later on.)

CHANGE AGENT #2: Infrastructure and Economic Interdependencies

The second significant change agent for the modern world involves the often-invisible backbone of modern commerce: the loosely but undeniably interconnected network of systems that control everything from the movement of goods and people to the transfer of money and the flow of electricity. The scale of the interconnectedness becomes more apparent when one considers that high speed, international communications networks are the backbone of modern commerce, and that they rely heavily on host nation power supplies, which in turn rely on other critical nodes, including the roads that employees use to get to work and the freight delivery network which delivers needed fuel and replacement parts. This is why, according to a report from Deloitte Research, "Disruptions in the supply chain...can have consequences far beyond their immediate geographic vicinity. A snowstorm in the Alps can affect train lines all throughout Europe. The closure of the ports in Hong Kong and Singapore, which together process more than one million containers a month, for just a few days, would wreak havoc on global trade."[11] Similarly, the interstate commerce system relies on safe and open highways and on a dependable nationwide supply of automobile fuel, our economy and food supply require open and safe borders and ports, and severe weather or other events can cause shortages in the availability of all of these goods.

Infrastructure and Economic Interdependencies

A second, less obvious but equally significant factor plays a major role in the functioning of our modern world. That factor is a direct product of the system itself, whereby a seemingly endless quest for maximum efficiency has created such lean systems that very little excess capacity remains. It is true that such efficiency confers important economic advantages in periods of normalcy by reducing waste and driving up profits. However, whenever there is a shock to the system, the lack of excess capacity in spare parts, cargo transportation, and even finished goods leaves the modern, streamlined firm without the ability to conduct business. This shock can come in the form of any disruptive event, be it severe weather, civil war, terrorism, or even just a labor dispute at any single critical node. And that shock, like waves crashing forth from a sudden tsunami, will spread inexorably across the global economic system in unforeseeable ways. As a result, we find ourselves collectively at much greater risk from what would once have been relatively minor disruptions with minimal impact on the overall system.

The term *critical infrastructure* refers to the interrelated economic sectors that are essential to "the minimum operations of the economy and the government."[12] Over the past two decades, there has been a growing awareness of the potential for catastrophic systemic failure growing out of such increasingly interrelated control systems. According to the George Mason University Critical Infrastructure Protection Project, the 1995 bombing of the Alfred P. Murrah building in Oklahoma City helped focus federal attention on the inherent vulnerabilities in our highly interdependent economic model. The attack killed 168 people and destroyed a single building. This one act, however, created a ripple effect across multiple agencies by destroying a regional payroll center, all the specialized investigative case files stored at an on-site federal records repository, and more. Soon thereafter,

Key Infrastructures

- Power lines
- Water lines
- Gas lines
- Rail lines

- Highways
- Airports
- Seaports
- Phone lines

both the government and the private sector began to focus on the enormous risks associated with another potentially massive infrastructure failure—the computer-programming problems known as year 2000 compliance, or Y2K.[13]

The federal government responded to this growing concern about interdependent critical infrastructures when President Clinton issued Executive Order 13010, creating the Presidential Commission on Critical Infrastructure Protection (PCCIP) to explore the overlap between critical infrastructures and national security.

Today, despite more than a decade of effort and the additional attention paid to such concerns following the terrorist attacks of September 11th, our collective understanding of the interdependencies among, for example, the railway, port, communications and energy industries remains elementary. Consequently the next significant event will likely have far-reaching but unforeseen consequences. In the international realm this situation is far worse, because even fewer resources have been allocated for learning about the potential for cascading impacts from the failure of any single node, and poorer nations have less ability to rapidly reconstruct their incapacitated infrastructure.

CHANGE AGENT #3: Discontinuous Events

The third Change Agent is the advent and prevalence of what may be called *discontinuous events*, meaning those events that are significant enough to create disruptions to the normal flow of commerce, transport, and the economic system. It is not that such events are new, for severe weather, war, famine, and attacks from fringe elements of society have long plagued the world. What is new here is the degree to which the impact of any otherwise disconnected event can reverberate throughout the global system as a whole. For this reason, such events must be considered in a new light.

In the post-9/11 world the most obvious and perhaps the most significant type of discontinuous event is a terrorist attack—for the same forces of globalization that empower individuals and small businesses to compete economically with more established firms have given even small terrorist groups a disproportionately powerful ability to disrupt the world around them. Not only does the internet allow such groups to spread their message,

Discontinuous Events

A third factor is what could be called *discontinuous events*, which refers to those disruptive occurrences that alter normal patterns and cause changes in the availability of services and the free flow of goods. In addition to the obvious category of severe weather and extreme political upheaval, severe discontinuities can be brought about by terrorists and other criminal elements. These activities are tied to globalization because, as an outgrowth of the market forces that are driving the world to greater integration, certain significant segments of the world's populace have voiced violent opposition. This opposition is embodied most notably in both the terrorist threats of those seeking to roll back modernity and the less severe but still important anti-globalization movements that continue to oppose these trends of global integration, often through violence. This "rejectionism" is a significant factor in the increased risk facing all manner of players in the international economic marketplace. Thus, it deserves to be examined, along with other traditional causes of discontinuity, such as severe weather and civil war.

The same forces of globalization that empower the individuals and small businesses to compete economically with more established firms have given even small terrorist groups a disproportionately powerful ability to disrupt the world around them. It is not that such events are new, for severe weather, war, famine, and attacks from fringe elements of society have long plagued the world. What is new is the degree to which the impact of any otherwise disconnected event can reverberate throughout the global system as a whole.

recruit new adherents, and coordinate their actions, but it also serves as an easy and reliable repository for their manuals regarding recruiting, planning attacks, and building bombs. As a result, the paramilitary experience, skills, and technologies that enable terrorists and insurgents to use improvised explosive devices and car and truck bombs to such effect are being spread far and wide. In addition, more and more detailed information about trade routes and schedules, building diagrams, and day-to-day corporate business dealings can be found on the internet and used by potential terrorists or criminals.

One of the most powerful root causes of many modern Islamic Fundamentalist terrorist movements is globalization itself. This is because the hegemonic power and dynamism of the economic forces that have driven integration have sown the seeds of rejectionism for various sects, including most significantly the "jihadist" terrorist movements emanating from Saudi Arabia, Yemen, Egypt, and elsewhere in the Middle East.[14] The repercussions of these anti-modernity elements can be felt nearly as widely as the forces of modernity and globalization themselves, for they use the very same technologies, integration of transportation, and networks to carry out their destructive activities.

Al Qaeda's primary target on 9/11 was not just any tall building, but the World Trade Center. Osama Bin Laden also has openly declared his intent to target the American economy, especially oil production and financial infrastructure. Therefore, the future could well bear witness to a prolonged period of select attacks that adversely impact the value chains of firms, by significantly impeding the free flow of goods and services.[15]

Recent history also offers a wealth of evidence about disruptive effects that are caused by nature, not man, such as the Indian Ocean tsunami of December 2004. Indeed, one full year after the disaster, news outlets report that the tsunami might have resulted in more than 200,000 people dead or missing across eleven countries.[16] Hurricane Katrina, which devastated New Orleans in August of 2005, is estimated to have cost at least 1,300 lives and more than $100 billion in damages.[17] The impacts of such events are magnified many times over by their long term impact on a nation's economic productivity, its currency and the repayment of international loans, and more tangible aspects, such as the cost of oil and other goods and services whose delivery have been disrupted.

Overview of Total Security Management

TSM has become important as an outgrowth of two parallel realities. First, due to globalization the world is an inherently more interconnected and interdependent place. Second, just as market economic forces are driving the economy to newly efficient, but also fragile, means of production and delivery of services and goods, the market can, should, and must learn to reward those firms that prudently manage their risks by being better prepared to deal with discontinuous events. Changes in

Total Security Management (TSM)

Total Security Management is the business practice of developing and implementing comprehensive risk management and security practices for a firm's entire value chain. This includes an evaluation of the preparedness of suppliers, distribution channels, internal policies, and procedures for discontinuous events such as terrorism, political upheaval, natural disasters, and significant accidents.

FIGURE 1.1   The core principles of TSM.

the way individuals and organizations perceive certain aspects of business are a logical outgrowth of the systemic impact of the three change agents discussed previously. One significant aspect of this change is a redefinition of security as it pertains to the firm's well being. At its core, Total Security Management provides a framework for identifying, implementing, and rewarding effective security practices. TSM is also a unifying theory that focuses upon a critical success criteria found throughout the business world—namely the security and resiliency of the enterprise. The TSM approach creates value through rewarding effective security-related business practices, the entire philosophy being predicated on the belief that security in the broadest sense is an integral part of all the business processes of all firms.

Traditional Security Versus Total Security

It is important to note that TSM does not use the term *security* in the narrow sense of the physical protection of gates, cyber security and network passwords, or firewalls and anti-virus software. This distinction is not made to minimize the importance of such measures. Having appropriate physical and cyber security for the purpose of monitoring and acting as a point defense against individuals seeking to do harm is critical to any sound and secure business.

But there is a broader approach that can be taken, which includes *Total Security* as a means for creating more value by inserting appropriate security considerations into the process for matters such as selecting corporate value chain partners, making determinations about certain vendors, and assessing internal and external resiliency and disaster preparedness factors.

For example, if a company takes on the task of redesigning certain access controls for security reasons, it can often make a case for improving overall business processes through the introduction of technology that will automate current manual processes. Technology, if properly applied, can reduce labor costs and liability concerns by removing a person from some potentially dangerous task, which in turn reduces insurance premiums and ensures better control and tracking over the people and vehicles coming in or out of a facility. TSM is a framework for the discovery of such dual-benefit solutions.

The Intellectual Foundation of TSM

TSM's intellectual foundations draw heavily from the seminal work of Dr. W. Edwards Deming, the U.S. government statistician who espoused *Total Quality Management (TQM)* as a foundational corporate philosophy of continuous quality improvements for manufacturers. TQM has been credited with successfully empowering and improving firms on multiple continents, notably beginning with Japan's resurgent economic success throughout the post-World War II period. Later, as a period of malaise hit the U.S. economy, TQM became a powerful force for change at major firms such as Ford, Chrysler, Westinghouse, Harley-Davidson, and Motorola. More recently, TQM, and a related derivative called "Six Sigma," has been showcased at such firms as AT&T, IBM, and Xerox.[18] As several transportation and supply management experts have noted, the lessons of TQM apply directly to today's challenges of securing transportation networks and creating appropriately resilient international value chains.[19] According to Professors Lee and Whang of Stanford University Business School, "The central theme of the quality movement—that higher quality can be attained at lower cost by proper management and process design—is also applicable in supply chain security."[20]

Accordingly, by developing the right approaches, harnessing new technologies, and re-engineering security processes, firms can attain better value chain security at a lower cost.

Echoing total quality's enterprise-wide approach, Total Security is predicated on the belief that good security must become an enterprise-wide effort. Under TQM, each person in the firm must be empowered to look for weaknesses in the status quo, to challenge the assumptions of the current system, and to create new and more resilient processes for accomplishing the mission of the enterprise. This is why TQM takes a holistic approach, applying basic bedrock principles across all levels, from the lowest-paid workers right up to the CEO. This effort to provide ownership over everything the firm produces makes creating quality a paramount task for each individual in the organization. In so doing, it empowers all of the firm's people to foster improvement throughout the organization. Indeed, one of the most significant aspects of TQM is refocusing the firm's energy—making certain that their employees are not merely doing their respective jobs, but are helped and enabled by management to do their jobs *better*, though proper education and support.

One idea for security improvement in keeping with the TQM approach is the redesign of how supply chains are secured, in terms of the actual containers being inspected at various points along the path, from field to factory to warehouse to retailer. At present, the supply chain system primarily uses spot-checks to verify compliance with bills of lading and other regulatory schemes at various control points. This is similar to how, before TQM, manufacturers would attempt to minimize variance by spot-checking final goods for quality assurance. Dr. Deming discovered that such an approach inherently accepts processes that create unacceptable inefficiencies because of the variability of quality designed into the process. It is also expensive because it disrupts the flow of finished goods headed to customers. As Lee and Whang write, " . . . in manufacturing the way to eliminate inspections is to design and build in quality from the start. For supply security, the analogy is to design and apply processes that prevent tampering with a container before and during the transportation process."[21]

Therefore, TQM would suggest that the solution to concerns with supply chain security is to redesign the packaging and handling process, as opposed to increasing inspections, which creates sinefficient costs in terms of bottlenecks, unpredictable delivery times, and direct economic losses such as demurrage.

The TQM approach includes a strong emphasis on the activities and processes of a firm's value chain partners, which can have important implications for forming lasting relationships with suppliers, vendors, and so on. The report *Effective Practice in Business Continuity Planning for Purchasing and Supply Management* states, "With a focus and philosophy on quality, a buyer and supplier can engage in a close relationship that results in continually improving inter-organizational processes, which therefore reduces the chance of process failures that result in product flow stoppages."[22] In today's age of outsourcing and increased specialization and interaction all along the value chain, the parallel for security is clear: TSM must incorporate an understanding of the risk and resiliency factors that affect an organization's entire value chain and security processes, and this understanding must be harmonized all along the value chain to result in better security for all parties.

Finally, TQM stresses the importance of creating definable, measurable goals. This focus on the *quantitative* aspects of good management is an important one, because it leads to the creation of measurable improvement and enables people to demonstrate whether or not their actions and efforts have created increased value for the firm. At the same time, it enables people outside of the firm to understand where the firm is headed and to measure its progress, which in turn leads to tangible rewards from evaluators such as investment analysts, insurance companies, and shareholders.

Total Security Management is an opportunity to differentiate a firm from others in the market space by reducing the probability of being affected by any disruptive event.

The same process must be created for today's global transportation firms, whereby measurable efforts and sound investments that are in accordance with best practices for survivability and resiliency in the face of disruptive shocks to the system can be recognized and valued as such.[23] Private sector businesses must embrace Total Security Management as an opportunity to differentiate their firm from others in the market. By reducing the probability of disruptions related to discontinuous events aimed at a firm's facilities, industry, or value chain, businesses can create value by initiating policies and procedures that anticipate threats posed by nature, negligence, or terrorism.

Dr. W. Edwards Deming's Fourteen Points for the Transformation of Management:*

1. Create constancy of purpose toward improvement of product and service, with the aim of becoming competitive, staying in business, and providing jobs.
2. Adopt the new philosophy.
3. Cease dependence on inspection to achieve quality. Eliminate the need for inspection on a mass basis by building quality into the product in the first place.
4. End the practice of awarding business based on the price tag. Instead, minimize total cost. Move toward a single supplier for any one item, on a long-term relationship of loyalty and trust.
5. Improve consistently and forever the system of production and service, to improve quality and productivity, and thus constantly decrease cost.
6. Institute training on the job.
7. Institute leadership. The aim of supervision should be to help people and machines and gadgets do a better job. Supervision of management is in need of an overhaul, as well as supervision of workers.
8. Drive out fear, so that everyone may work effectively for the company.

9. Break down barriers between departments. People in research, design, sales, and production must work as a team, to foresee problems of production and in use that may be encountered with the product or service.
10. Eliminate slogans, exhortations, and targets for the workforce asking for zero defects and new levels of productivity. Such exhortations only create adversarial relationships, as the bulk of the causes of low quality and low productivity belong to the system and thus lie beyond the power of the workforce.
11. Eliminate quotas. Eliminate management by numbers and numerical goals. Substitute leadership.
12. Remove barriers that rob the hourly worker of his right to pride of workmanship. The responsibility of supervisors must be changed from sheer numbers to quality.
13. Institute a vigorous program of education and self-improvement.
14. Put everybody in the company to work to accomplish the transformation. The transformation is everybody's job.

*Adapted from the website of the W. Edwards Deming Institute, available at http://www.deming.org, and originally presented in W. Edwards Deming's 1982 book *Out of the Crisis*. Reprinted with permission.

TSM is Optimized Where Security Meets Efficiency

TSM decisions should also be guided by the firm's need to ensure implementation of dual-benefit solutions whenever possible. Because value chain *security* is separate from but inherently tied to value chain *efficiency*, TSM helps a company focus primarily on solutions that overlap. This region of overlap includes such areas as regulatory compliance, theft prevention, safety, and value chain visibility, as well as firm-specific requirements such as selection of vendors.

Risk Management and The Case For TSM

The traditional business approach to handling operational risk is termed *risk management*, which is itself comprised of

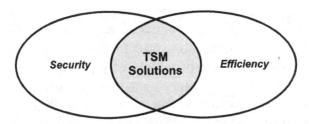

FIGURE 1.2   TSM is optimized where security meets efficiency.

three distinct tasks: measurement of risks, selection of risk mitigation alternatives, and acceptance of residual risk. In this approach, it is essential that firms properly examine and quantify the full spectrum of risks they face, for an error in the initial calculations will mean improper decisions in subsequent determinations of risk mitigation versus acceptance. This calculation is no easy task, for the broader security picture must be kept in mind when determining the optimal course of action—not only over the short term, but the medium and long term as well. The Total Security approach highlights additional factors for consideration and evaluation that might otherwise be missed.

Firms must properly examine and quantify the full spectrum of risks they face, for an error in the initial calculations will mean improper decisions in subsequent determinations of risk mitigation versus acceptance.

This approach does not argue against outsourcing *per se* or negate the need to reduce operating costs by selecting low cost vendors; rather, it simply brings to light certain exogenous factors that, if unmeasured, can lead to an improper allocation of risk.

The first step in risk management is to study current risks and various mitigation alternatives, but to do so while using business imperatives to find ways where security complements, but does not undercut, corporate business practices. At the same time, the balance between normal business functions and the potential deterrent value of appropriate security is a key to sustainable business practices. Over time the marketplace should differentiate among firms based on their resilience and disaster preparedness, and the firms that act now will be ahead of the curve in reaping the rewards of this shift in investor and consumer confidence.

As a case in point, consider the international financial services industry, many major components of which outsource a significant quantity of daily operations to Bangalore, India. This is an exceptionally cost-effective practice for daily operations because Bangalore offers literate, skilled, and dedicated employees who work at wages dramatically lower than are available in more developed nations. However, the entire industry is at risk if there is a nuclear confrontation between Pakistan and India, and, though this is generally considered a remote possibility, as recently as June of 2002 many governments, including the U.S., recalled employees stationed in both countries based on the very real fear that tensions would escalate into violence and potentially a nuclear exchange. Fortunately, tensions were eased in that instance, but those in the intelligence community assessed war at that time to be a very real possibility.[24] Today, with Pakistan's President General Pervez Musharraf continuing to face extreme internal pressures and having survived multiple assassination attempts in the past three years, concerns about India's well-being cannot be dismissed lightly.

Similarly, consider an insurance firm that decides to outsource certain financial accounting procedures to a developing nation, but fails to account for the reality that severe seasonal weather in that nation periodically disrupts communications during a critical operations period for that industry, such as the time of year when the majority of existing policies are either renewed or renegotiated.

Even if on average such disruptions only occur every few years, floods or storms that cause communications outages during a significant time period ultimately could cost the firm more than its saves from moving the processes overseas. Though the firm may reduce daily operational costs through outsourcing, it has (knowingly or unknowingly) accepted additional risks that may, in the worst case scenario, end up costing the firm more than other similar courses of action. Ultimately, overlooking or discounting such a significant risk factor exposes the firm to unmeasured risk.

In addition, the advent of global terrorism, combined with the increasing interconnectedness of the systems which we rely upon to run, monitor, and respond to nearly every activity of our daily lives, has given rise to tremendous inherent risk and uncertainty. This risk extends to all types of businesses, and even recognized industry leaders often remain surprisingly unprepared to prevent or respond to disruptive events. Determining how much risk to accept, while reasonably mitigating other risks and continuing to carry out normal business operations, can be a difficult calculation to make.

Because the inexorable effects of failures in the interconnected, interdependent infrastructures and global economic system are not readily visible, the second, third, and fourth order effects of any failure will surely surmount the remediation capabilities of most governments around the world. As the people of Indonesia learned following the December 2004 tsunami, the people of Pakistan learned following a deadly earthquake in October 2005, and even Americans learned when Hurricane Katrina devastated New Orleans in August 2005, major events can overwhelm the response and recovery capacity of most, if not all, governments. Given this reality, and the fact that firms in every nation are increasingly reliant upon the presumed safe, secure operating conditions of their affiliates and partners across the globe, it is incumbent upon each firm to reexamine their exposure relative to potential disruptions.

The ability to measure such risks and to make better risk management decisions is a critical component of the TSM process, because it demonstrates that security process improvements are not a net loss and that proper security can create value for the firm. In other words, the Total Security Management approach is the *sine qua non* of business continuity planning and the foundation of rational risk management. Determining how much security is enough is a function of a tried and true risk assessment process, whereby various threats are identified, assets values are determined, and risk mitigation alternatives are considered.

In fact, if properly assessed and implemented they become a critical node in the system of corporate checks and balances, as well as a legitimate means to avoid the possible imposition of costly (and often ineffectual) government regulations.

Conclusion

The continuing processes of change that are redefining global business practices and realigning global economic power have created the need for a Total Security Management approach to enterprise resiliency. Three distinct but related change agents embody the most significant aspects of this process, and they are a) globalization, b) infrastructure and economic interdependencies, and c) discontinuous events. Total Security Management, which is at its core a unifying theory that draws in and seeks to build value from the totality of today's security and continuity of operations imperatives, has become a required foundation for sound business practices because of the net effect its implementation can have upon a firm's overall value.

The TSM approach to global economic evolution is a reaction to the reality that our increasingly connected world is also increasingly fragile. This is true not only because deep economic integration makes each of our value chains intertwined in unexplored and often unexpected ways, but even more fundamentally because the forces of free and open competition are driving production costs to their most basic level, and in turn reducing any perceived ability to maintain excess capacities of nearly any kind. The inherent weaknesses in this just-in-time world have been laid bare in the ripple effects and cascading impacts of supply chain disruptions witnessed in recent years following terrorist attacks, severe weather events, and even transportation labor disputes. As Deloitte Research observes in *Prospering in the Secure Economy*, "This new period is defined by greater vulnerability, increased threat awareness, regulatory compliance, and rapid response to change. It is not an environment signified by the global war on terrorism, but of enhanced visibility—and responsibility—across supply chains and cross-border relationships."[25]

The reason Total Security Management is so critical in today's world is that it is predicated on sound business and security practices that ensure preparedness for all manner of disruptive events, whatever the cause. It is a system that recognizes the challenges created by the forces of globalization and embraces them as opportunities for process improvements. At the same time, TSM justifies investments in an overall resiliency that ensures that a firm is optimally positioned to reap the benefits of medium- and long-term mission surety. Value can be created by avoiding single points of failure in the global value chain and by ensuring an ability to rapidly resume operations should a disruption take place. Finally, TSM asserts that knowledge of risks leads to better decisions about risk mitigation versus risk acceptance. If properly measured, risk management initiatives can in turn create value for the firm through its use as a market differentiator that initiates a reward response from consumers, investors, and analysts.

Notes

1. This final portion of this passage refers to the very public debate surrounding the proposed Dubai Ports World takeover of operations at several U.S. ports following the approximately $7 billion purchase of a British port operating firm.
2. Deloitte Research, *Prospering in a Secure Economy*, 2004, <http://www .deloitte.com/dtt/cda/doc/content/DTT_DR_ProsSecFull_Sept2004 .pdf#search='Deloitte%20Research%20prospering%20in%20the%20se cure%20economy> (11 March 2006).
3. IBM Center for the Business of Government, *Enhancing Security Throughout the Supply Chain*, April 2004, <http://www .businessofgovernment.org/pdfs/Closs_Report.pdf> (20 February 2006).
4. LCP Consulting and the Centre for Logistics and Supply Management, Cranfield School of Management, *Understanding Supply Chain Risk: A Self-Assessment Workbook*, 2003, <http://www.som.cranfield.ac.uk/som/ research/centres/lscm/downloads/60599WOR.PDF> (8 April 2006).
5. Stanford University Global Supply Chain Management Forum, "Innovators in Supply Chain Security: Better Security Drives Business Value," The Manufacturing Institute, July 2006, http://www.nam.org/ supplychainsecurity.
6. Thomas L. Friedman, *The World is Flat* (New York: Farrar, Straus, and Girous, 2005), 12-13.

7. Stan Gibson, "HP Clones P&G's Cincinnati Accounting Ops in India," E-Week.com on the Web, 19 March 2006, <http://www.eweek.com/ article2/0,1895,1939545,00.asp> (25 April 2006).

8. GlobalisationGuide.org, *When did globalisation begin?*, 2002, <http:// www.globalisationguide.org/index.htm> (8 December 2005).

9. GlobalisationGuide.org, *When did globalisation begin?*, 2002, <http:// www.globalisationguide.org/index.htm> (8 December 2005).

10. Thomas L. Friedman, *The World is Flat* (New York: Farrar, Straus, and Giroux, 2005), 111.

11. Deloitte Research, *Prospering in a Secure Economy*, 2004, <http://www .deloitte.com/dtt/cda/doc/content/DTT_DR_ProsSecFull_Sept2004 .pdf#search='Deloitte%20Research%20prospering%20in%20the%20se cure%20economy> (11 March 2006).

12. Bill Clinton, "Critical Infrastructure Protection," Presidential Decision Directive 63, 22 May 1998, <http://www.fas.org/irp/offdocs/pdd/pdd-63 .htm> (20 December 2005).

13. George Mason University, "Oral History Project," The Critical Infrastructure Protection (CIP), <http://chnm.gmu.edu/cipdigitalarchive/> (22 April 2006).

14. Thomas L. Friedman, *The World is Flat* (New York: Farrar, Straus, and Giroux, 2005), 391-392.

15. G. Gordon Liddy et al., *Fight Back* : Tackling Terrorism, Liddy Style (New York: St. Martin's Press, 2006), 27-28.

16. Ann O'Neill, "A year of epic disasters, terrorism and politics," *CNN on the Web*, 2006, <http://www.cnn.com/2005/US/12/31/2005.review/ index.html> (1 April 2006).

17. Ann O'Neill, "A year of epic disasters, terrorism and politics," *CNN on the Web*, 2006, <http://www.cnn.com/2005/US/12/31/2005.review/ index.html> (1 April 2006).

18. Wikipedia, *Total Quality Management*, <http://en.wikipedia.org/wiki/ Total_quality_management> (13 April 2006).

19. Yossi Sheffi, *The Resilient Enterprise* (Massachusetts: The MIT Press, 2005), 132. *See also* Hau L. Lee and Seungjin Whang, *Higher Supply Chain Security with Lower Cost: Lessons from Total Quality Management*, October 2003, <https://gsbapps .stanford.edu/researchpapers/detail1.asp?Document_ID=2273> (20 December 2005).

20. Hau L. Lee and Seungjin Whang, *Higher Supply Chain Security with Lower Cost: Lessons from Total Quality Management*, October 2003, <https://gsbapps.stanford.edu/researchpapers/detail1.asp?Document_ ID=2273> (20 December 2005).

21. Hau L. Lee and Seungjin Whang, *Higher Supply Chain Security with Lower Cost: Lessons from Total Quality Management*, October 2003, <https://gsbapps.stanford.edu/researchpapers/detail1.asp?Document_ ID=2273> (20 December 2005).

22. George A. Zsidisin, Ph.D. and Gary L. Ragatz, Ph.D. and Steven A. Melnyk, Ph.D., "Effective Practices in Business Continuity Planning for Purchasing and Supply Management," 21 July 2003, <http://www.bus.msu.edu/msc/documents/AT&T%20full%20paper.pdf> (1 February 2006).

23. Yossi Sheffi, *The Resilient Enterprise* (Massachusetts: The MIT Press, 2005), 132-133.

24. Margaret Warner and Ambassador Lalit Mansingh, "PBS Online NewsHour: The Indian View," 5 June 2002, personal interview, <http://www.pbs.org/newshour/bb/asia/jan-june02/india_6-5.html> (27 April 2006).

25. Deloitte Research, *Prospering in a Secure Economy*, 2004, <http://www.deloitte.com/dtt/cda/doc/content/DTT_DR_ProsSecFull_Sept2004.pdf#search='Deloitte%20Research%20prospering%20in%20the%20secure%20economy> (11 March 2006).

"*Everybody has accepted by now that change is unavoidable. But that still implies that change is like death and taxes it should be postponed as long as possible and no change would be vastly preferable. But in a period of upheaval, such as the one we are living in, change is the norm.*"

—Peter F. Drucker, *American management expert*

Chapter Two

The Total Security Management Framework

The three change agents detailed in the previous chapter drive us towards an understanding of how the world is changing and the reasons that Total Security Management is necessary.

Yet, if one accepts the significance of the three change agents, why is it that the effects are not felt evenly across all firms, as evidenced by the fact that certain firms fail while others succeed? One of the reasons is simple, but perhaps not obvious: the firms that currently succeed by applying appropriate security and resiliency practices use different terminology to describe their approach. Nonetheless, in whole or in part, they exemplify the best of what the TSM approach has to offer. As a result, these firms are able to thrive even under the pressures of globalization, confidently withstanding complications that can arise from critical infrastructure dependencies, and respond quickly and adeptly to disruptive events.

An analysis of the factors that make certain firms more resilient was undertaken in 2003 in a report funded by AT&T, entitled *"Effective Practices in Business Continuity Planning for Purchasing and Supply Management."*[1] The authors concluded that two issues are at the heart of a company's current value chain management problems:

· Reduced visibility into the firm's detailed value chain because of increasing complexity at all levels
· Reduced control and an inability to influence the actions of key suppliers, especially those one or two levels removed from the firm through supplier networks

Specifically, the study states that, "These problems are compounded by the application of "lean" practices, which reduce supply-chain buffers in the form of inventory, lead-time, and capacity. Consequently, many managers are unaware of what their suppliers are doing to ensure business continuity. With so many companies depending on a reduced set of suppliers for key components, the potential for problems is real and increasing. If a key supplier is unable to perform, the impact on the firm—as measured financially, strategically, and in terms of market share—can be sizable, and even catastrophic."[2]

Another report generated in the United Kingdom, similarly concluded that, "The single-minded pursuit of efficiency within supply chains (has) inadvertently increased the vulnerability of those same supply chains to unforeseen disruptions."[3] This realization that the lean, distributed supply chains, combined with ever-expanding outsourcing of various specialized tasks and the prevalence of risk created by the three change agents, presents a new era in value chain management and is the driving force behind the development of Total Security Management.

Fortunately, firms have an opportunity to turn these challenges into rewards by creating value through the application of effective security practices. According to Deloitte Research, firms must master certain challenges to succeed in the new business environment. These include assessing and managing threats specific to the company, strengthening crisis management capabilities, implementing an enterprise-wide security plan, addressing end-to-end supply chain security, and maximizing shareholder value through investments in security.[4]

Achieving a Total Security Management solution set requires a comprehensive action plan based on understanding existing and emerging vulnerabilities in a new way. It requires the teamwork and active participation of multiple players throughout the firm's value chain. And at the end of the day, it requires the adoption and application of new tools to understand and implement risk reduction and resiliency-enhancing strategies. The following section describes some of these key tools in detail, and provides the transportation and security professional with the necessary resources to create value beginning with TSM's Strategic Pillars and Operational Enablers.

TSM's holistic, enterprise-wide approach requires analysis and action that will achieve precisely these kinds of solutions by addressing the following three interlocking levels:

- *Strategic*: This is the long-term, big picture frame of reference that takes the long view of overall operations and sets strategy and goals for the firm's future in accordance with the Five Strategic Pillars (as defined below).
- *Operational*: This covers the overall issues of how the company is run, division-by-division, its interactions with various members of the value chain, and in accordance with the Four Operational Enablers (as defined later in the chapter).
- *Tactical*: These are detailed specifics for how an individual firm should apply core TSM principles and the tenets of the TSM Value Creation Model to the firm's daily operations, including the scope of the initial security assessment, a list of who is on the implementation team, the timeline for deliverables, and the metrics for assessing results. Each firm will define the tactical considerations individually, drawing from the TSM Value Creation Model (as defined later in the chapter).

FIGURE 2.1   TSM is an enterprise-wide approach

The Five Strategic Pillars

Total Security Management practices begin with adherence to a certain set of bedrock principles known as the "Five Pillars of TSM." Using these principles as a guide for implementing the processes and solutions makes it possible to create value for a firm, while also ensuring standardization across all entities that adhere to TSM. They are necessarily generalized because they serve as the overarching guide for TSM implementation— marking the left and right limits of the security processes, procedures and initiatives that need to be managed and the areas that need to be addressed for proper TSM implementation.

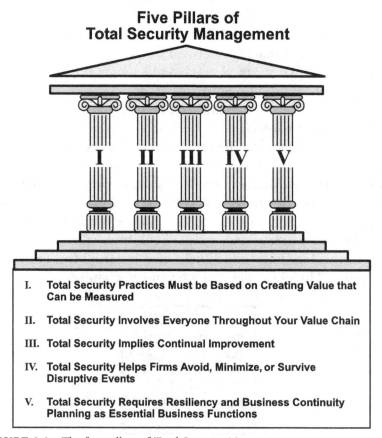

Five Pillars of Total Security Management

I.   Total Security Practices Must be Based on Creating Value that Can be Measured

II. Total Security Involves Everyone Throughout Your Value Chain

III. Total Security Implies Continual Improvement

IV. Total Security Helps Firms Avoid, Minimize, or Survive Disruptive Events

V.   Total Security Requires Resiliency and Business Continuity Planning as Essential Business Functions

FIGURE 2.2   The five pillars of Total Security Management

PILLAR #1: Total Security practices must be based on creating value that can be measured.

One of the most important rules in business is that in order to succeed, a firm must strive to create value in all of its activities. Total Security Management is no exception; indeed, the principle that value can be created through security actions is the most central element in the concept of TSM.

This value must be created both by the additional crisis readiness and the resiliency afforded to a firm that may be subjected to a significant discontinuous event and/or from process improvements and efficiencies found within the daily security practices themselves.

For example, solutions that redesign point-of-origin packaging practices to better ensure tamper-proof shipping and theft prevention can reduce the need for certain inspections, which in turn can save costs in terms of the time it takes to get the goods from the supplier to the end-user. Similarly, solutions such as the automated tracking of packages and transportation assets through radio frequency identification (RFID) tags or satellite communications to ensure better value chain visibility can contribute to these same objectives. Such solutions have the potential to enhance security while reducing certain labor costs associated with manual tracking processes, and will ultimately be more sustainable than practices that don't have savings to offset their implementation.

TSM practices also take into consideration the context of how the firm operates in devising security processes. Consider the competing concerns of businesses that maintain large amounts of cash on hand, such as banks, amusement parks, and casinos. For such industries, a longstanding joke among security professionals has been that the best way to ensure one hundred percent security is to shut the doors and cease operations. While technically true, it is also possible for a firm to secure themselves out of business.

Accordingly, such businesses have invested heavily in ways to maintain an open flow for their customers along with a sense of security from appropriate combinations of surveillance equipment, safes, security guards, and insurance policies. Indeed,

Las Vegas in particular has implemented significant crime prevention protocols while ensuring the public's freedom of movement.

The right solution for creating sufficient value from various practices is also dictated by a firm's relative market position. Consider three hypothetical companies of three different sizes, all of which rely heavily on a distributed transportation network where both the cost of freight and its timely delivery greatly affect profitability.

Company A, which is a large company, may find that their best solution is to invest in creating a local transportation presence in a remote location. The potential value in this choice has to do with the fact that there is likely to be greater opportunity for growth and this new subsidiary entity will be able to create value by supporting the freight needs not only of the parent firm, but other firms as well, selling their excess capacity to recoup the cost of the infrastructure growth.

For Company B, a medium-size company, the better solution might be to band together with a consortium of other mid-size firms and seek both economies of scale and mutual protection against disruption through shared, distributed global transportation capacity.

Finally, Company C, a small firm, may find that contracting for the transportation services of Company A's subsidiary gives it the proper balance of highly modern and reliable transportation without the attendant costs of ownership.

It is important to understand, however, that there can be no one-size-fits-all answer to TSM. Total Security Management practices need to create value and show relevance to the functioning of the enterprise by playing to its strengths. This is why TSM is a process requiring self-evaluation, careful planning and integration, and thoughtful implementation, as opposed to merely a standard solution set that can simply be added to the existing corporate mix.

Today's reality is that a disruption can strike any link in the chain. In the end, it matters little whether the failure occurred during an in-house or outsourced process.

The backbone of the TSM approach is this understanding of the power of making relevant choices for each firm, and how such behavior ensures that players in the marketplace who pursue TSM will be able to create value in unique yet often complementary ways. By using the tools of competitive analysis to examine strengths and weaknesses relative to certain categories of risk and to the market space as a whole, firms can create new value and opportunity in complete harmony with their overall business model. By pairing this analysis with the implementation of select best practices, firms have the opportunity to turn security costs into centers of value creation.

PILLAR #2: Total Security involves everyone throughout your value chain.

Today's reality is that a disruption can strike anywhere in the chain, be it a value chain partner's factory, part of the network of transportation services that move materials from the fields to the factory to the distribution centers, or the network of computer and communication systems that support modern value chain operations.[5] Accordingly, TSM requires a holistic approach because the integrated nature of today's value chain solutions mean that a failure of any of the processes that enable a firm to conduct business could put the firm at risk of significant compromise. In the end, it doesn't matter much whether the failure occurred during an in-house or outsourced process. This is why TSM works to identify and mitigate risks wherever they may exist throughout the relevant interlocking value chain network.

At the very least, TSM seeks to be able to knowledgeably account for the actual amount of risk being faced, in order to more properly balance other related strategic planning such as decisions affecting corporate growth or outsourcing.

There are additional benefits to be realized from more open and involved discussions with various partners, however. At the most basic level, the simple human bond of knowing the person in the other firm can be a powerful tie that helps smooth out issues big and small, in both times of harmony and times of crisis.

In addition, a firm's knowledge about the market conditions that affect its partners can provide better input to its own strategic decisions, including the early detection of relevant industry trends and comprehension of new conventional wisdom and related opportunities for both products and services development and targeted growth and investment.

PILLAR #3: Total Security implies continual improvement.

Like Total Quality Management, discussed in Chapter 1, TSM cannot be viewed in isolation or as if the process were a one-time effort that did not require follow-up. As with many best practices in a dynamic business environment, Total Security Management is a process that builds upon previous changes and makes incremental improvements all along the way. It requires active and concerted effort over time in order to achieve its full benefits.

This process of continual improvement can take many shapes, including a benchmarking against other firms in the industry, the testing of existing plans and procedures through realistic exercises, ensuring the adoption of appropriate emergent technologies for both security and process improvements, and seeking information about the changing nature of threats and weaknesses that are found not only within the industry but throughout the value chain. Because the competitive economic environment is continually changing and evolving, a firm's TSM implementation must be equally dynamic, and should be updated using feedback that fosters additional improvements and optimizes options for implementing best practices and workable solutions. Continual improvement has become more significant now that globalization is taking hold, and the competitive business environment demands change as the world's economies become increasingly interconnected and dynamic.

PILLAR #4: Total Security helps firms avoid, minimize, or survive disruptive events.

By striving for continual security and business process improvement, a firm can become better aware of various threats and more attuned to when things are happening that might indicate a potential disruption. Avoidance and minimization of disruptive events can save a firm money. In other words, value can be created and certain forms of risk can be mitigated merely by thoughtful reflection and a willingness to challenge status quo processes. Accordingly, TSM seeks solutions that not only help firms assure an ability to recover from those events that do occur, but also enable firms to avoid or minimize disruptions that might otherwise take place. This pillar is a critical component of enterprise survivability because it recognizes that no matter how well a security incident can be handled, a problem that is avoided works out even better.

PILLAR #5: Total Security requires resiliency and business continuity planning as essential business functions.

Although the detailed specifics of any future anomalous situation cannot be known *a priori*, it is possible to plan for and test against certain more likely types and classes of disruptions. Given the increasing risk and under-recognized systemic fragility into which Total Security Management was born, it is no surprise that it requires resiliency and business continuity planning for all significant aspects of a firm's business processes.

Indeed, this is all the more important if one's competitors are not well-prepared to handle the impact of a disruptive event, for what may mean a slight disruption to a firm with distributed risk and proper resiliency may be a major distraction or even unrecoverable event for another, less-prepared firm. The goal in TSM is to ensure that the enterprise is stable enough to appropriately withstand whatever challenges lie ahead. Even so, since businesses cannot avoid all risk of being impacted, the prudent course is to invest the resources necessary to provide for the response and recovery efforts that will be needed when the inevitable disruption does occur.

TSM's Four Operational Enablers

The four operational enablers provide the framework for translating the strategic concepts described in the Five Pillars into a direct application for analyzing a firm's specific TSM initiatives. Taken together, the four operational enablers represent the significant focus areas for efforts related to TSM and offer the broad outline of where to begin to create solutions that address security threats and vulnerabilities. They also describe the manner in which firms should directly address the specifics of the TSM Value Creation Model.

ENABLER #1: Best Practices Implementation

Best practices awareness and implementation is perhaps the most fundamental of the operational enablers because it requires a firm to both understand its own operational processes and policies and to reach out in order to assess the best practices that it may want to incorporate from throughout the industry. The areas to review include general security and business practices as well as vendor relations value chain management considerations, and of course, technological improvements that affect a firm and its suppliers.

Although this enabling element builds on the common business practice of measuring firms against the actions of their peers, it goes further than the usual analysis of just determining what others are doing so that the firm can follow suit. Indeed, one of the key tenets of implementing TSM is remembering that what others are doing is not always or automatically a best practice in terms of TSM. Initiatives implemented by one firm may entail hidden effects that create additional risks for another. The proper approach includes conducting a thorough assessment of all relevant trends, practices, and technologies and then selecting the right risk mitigation and process improvement initiatives to match the firm's prerogatives.

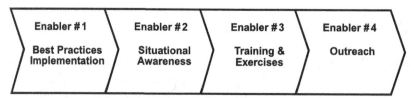

FIGURE 2.3   The four enablers of Total Security Management

ENABLER #2: Situational Awareness

Situational awareness has to do with collecting and processing information for the benefit of the decision makers who have to set the priorities for the firm. Situational awareness means using a systematic approach for dealing with and understanding a firm's current processes, so as to be better able to respond to disruptive events. In particular, the knowledge of what *is* and what *could be* leads to more optimal investment decisions with regard to new business practices and modernization efforts. More than ever before, security and resiliency considerations should play a role in such decisions.

Naturally, a firm must not only understand what is happening with its specific facilities, equipment, and personnel, but also be knowledgeable about similar details for its significant value chain partners. This is a fundamental tenet of all TSM implementation. This knowledge is especially critical when it comes to rapidly reacting to an impending or recent disruptive event—precisely the time when this information is likely to be the hardest to come by, which is why the TSM approach requires diligent, continuous work on situational awareness in advance of any disruptive events. In particular, understanding a firm's relationships to its various suppliers and related business partners and knowing how to track critical elements wherever they may be (known as "value chain visibility") are critical areas for situational awareness.

Many leadership and other executive training programs, including well-regarded programs at the University of Pennsylvania's Wharton Business School and the Harvard University Business School as well as courses taught by former New York Mayor Rudy Giuliani, emphasize the role of the individual leader in disaster recovery and business continuity. However, without the knowledge of both what *should* be happening and what actually *is* happening, even the most effective leader will be without recourse. When it comes to business continuity and/or recovering from a disruptive event, knowledge truly is power. It therefore follows that situational awareness is a key aspect of establishing and maintaining Total Security Management.

ENABLER #3: Training and Exercises

It is said in military circles that, "No plan survives contact with the enemy, so planning is nothing, but the process of planning is everything." This statement is also true for the preparation of dealing with disruptive events affecting the transportation industry, where the "enemy" may be natural or manmade, and where the effects must be dealt with under conditions of duress. It is for this reason that training events and exercises can often present the best value in terms of level of effort expended versus the dramatic gains in preparedness.

The first step, of course, is to ensure that all appropriate personnel receive any specialized training they may need to deal with threats to life and limb, as well as IT and related processes critical for business continuity. Next, a firm should develop an initial contingency plan so that all its employees have some fundamental concept of how to respond to such an event. But it is not enough to just have a plan, even one that most employees have read or that people are vaguely familiar with. Firms concerned with how they will *actually* react can only discover this by placing themselves in realistic training scenarios. It is only then that they will be able to identify some of the flawed assumptions underlying their plans, such as assuming that there will be working telephones or trucks that will be able to get fuel following a severe storm that affects overland transportation routes. Once recognized, these flaws can be remedied, so that on the day of the event there will be fewer surprises, maximizing the likelihood of a successful response.

ENABLER #4: Outreach

As concluded by Deloitte Research, "Companies that thrive in the secure economy will have security and sustainability built into their supply chains and greater security compatibility with their partners."[6] This is why the final operational enabler is that of outreach, referring to relations with both a firm's value chain and its other constituents. This is clearly an area of growing importance, given the increasing interconnectedness of our world.

The Three Types of Emergency Preparedness Exercises

- The least intrusive, though often quite instructive, format is called a 'tabletop' exercise, which, not surprisingly, often takes place inside around a large table. In this exercise, senior decision makers react to hypothetical events that are presumed to have occurred, while the fictionalized situation is altered according to a centrally-managed schedule. The intent here is to draw out the discussion of various events and the appropriate responses.
- The next format is called a "functional" exercise, wherein individuals within the organization and certain departments actually role-play against scripted scenarios, with some select sections of the firm actually carrying out portions of their response plan. This type of event is more involved than a tabletop exercise, but still allows for some degree of artificiality in terms of timing and reduces the requirements of manpower and resources that are involved for a full-scale exercise.
- The final format is called a full scale or "field" exercise, and is markedly more intricate, often requiring actual deployment of resources by multiple parties simultaneously. This, of course, requires the highest level of effort, but it also tends to highlight real-world problems that lesser-scale exercises might assume away. Though such exercises can often be run internally, there is a specific set of skills that have been built around the creation of successful exercises, and optimal implementation may require seeking additional outside support.

Understanding their partners' strengths and weaknesses can better enable firms to spot looming problems, as well as help them predict the potential disruptions that will be significant and those which may not amount to much at all. Getting the word out about security initiatives can hedge against a negative reaction following a disruptive event, and contributes to strengthening resilience throughout the network.

However, the basis for a firm's increased value chain interaction is relevant beyond just security concerns. Using the interaction and collaboration required to address mutually significant value chain challenges to build rapport can pay significant dividends over time. Opportunities to strengthen alliances and relationships, and to demonstrate and share information regarding areas of mutual interest and complementary core competencies, should be encouraged. Such openness can foster increased knowledge across a range of critical issues and increase efficiency in a variety of ways. For example, Wal-Mart has realized gains from fully integrating its suppliers into its sales cycle, allowing it to share near real time sales and inventory figures with a variety of partners throughout the world.[7]

Similarly, through the exercises discussed previously in Operational Enabler #3, a firm should also use outreach practices to involve its local and regional constituents, including law enforcement entities, in response planning procedures. By working with those who will be affected during the recovery process the firm can make alliances and create the goodwill which can be so essential during times of duress. Practices such as these can become assets when they're needed most, and, not surprisingly, the day of the adverse event is not the time to try to create those connections.

The need for outreach does not end with a firm's neighbors, however. As with so many TSM precepts, it extends to a company's entire community of value chain partners. The relationships and mutual dependencies created in advance of the event will become valuable resources in a firm's recovery processes, and this is especially important when it affects business across the globe and across different cultures.

Indeed, one of the more important opportunities to take advantage of is the potential to benefit from collective industry action. Although some industry-specific groups exist for collective action, such as Business Executives for National Security (BENS), the Air Transport Association (ATA), and the Association of Contingency Planners (ACP), their work to date in establishing truly comprehensive security policies has been limited, especially with regards to the transportation industry. Through outreach, a firm can lead the charge towards ensuring a common set of principles is adhered to, and also ensure that they have a voice in influencing certain regulatory activities in order to fight potential security regulations that are unduly onerous or which do not take business priorities into consideration.

Although TSM primarily looks to the private sector to solve its problems, through processes that reward economically sound investments in resilience and preparedness, working with government entities should be included in the outreach solution set. Building these relationships can represent a good way to help enable rapid recovery of operational capabilities. The fact remains that the governments of the world have the power to stop the flow of goods or otherwise affect global transportation in a myriad of ways. Mutual understanding and personal relationships can go a good deal of the way towards promoting mutual aid and easing the tensions that will accompany any significant disruption to the normal functioning of the global transportation system.

The TSM Value Creation Model

The most fundamental principal of TSM is that the process of managing risk and protecting core business assets is, if properly carried out, a means to create value. The five strategic pillars and the four operational enablers provide the philosophical underpinnings and the broad framework for TSM practices. It is in the application of the *TSM Value Creation Model*, however, where the "rubber meets the road," and the specific metrics for analyzing a firm's relative resiliency and preparedness for discontinuous events can be applied.

Transportation security initiatives that focus on security in the context of a firm's specific business imperatives have the most potential to create a real return on security investment, especially at those points where process improvement and security initiatives overlap.

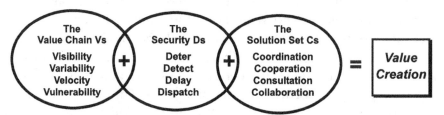

FIGURE 2.4 The TSM Value Creation Model

Transportation security initiatives that focus on security in the context of a firm's specific business imperatives have the most potential to create a real return on security investment, especially at those points where process improvement and security initiatives overlap. This recognition drives the TSM Value Creation Model, which is based upon (and shows the linkages among) the core elements of running a successful transportation business, defined here as the *Four Value Chain Vs*, *Four Security Ds*, and *Four Solution Set Cs*.

The Four Value Chain Vs

The TSM Value Creation Model begins with the nexus where business and security priorities overlap with respect to the value chain, a nexus that we identify as the Four Value Chain Vs. These include:

Visibility   The ability to identify movement in the value chain.

Visibility's importance on the business side of the equation is reasonably straightforward. Transportation is about movement and identifying movement is what gives a firm visibility into its value chain. The ability to identify this movement can represent a measure of success and differentiation from one's competitors where business processes are concerned, such as how FedEx famously revolutionized package delivery by providing clients the ability to know precisely where packages were and when they'd be delivered. In this manner, visibility represents one of many ways that a transportation firm can use visibility to create value and better compete in the marketplace.

The ability to identify movement in the value chain is also directly related to security, especially when combined with emergent technologies like RFID tags that can assist in reduction of theft, management of logistics for port and terminal operations, and more accurate verification of shipping manifests. Current conventional wisdom also suggests that "freight at rest is freight at risk." However, if a shipment can be verified to be moving through the value chain as scheduled, then certain assumptions can be made about the likelihood that the shipment has not been compromised. It is therefore reasonable to make a direct association between movement and security. Good value chain visibility also helps a firm to detect disruptions earlier, which in turn can provide value by giving them an edge in resolving minor disruptions before they grow out of control.

Variability   The level of consistency in the quality of value chain processes.

For decades the business world has embraced the fundamental tenets of Total Quality Management, including limiting or eliminating variability in the quality of products and processes. Indeed, controlling variability goes to the heart of the TQM philosophy that has enabled countless businesses to create value. Consistency and reliability—two concepts that are directly tied to variability—are key performance measures in transportation. Just as in other industries, variability is important in transportation, and firms go to great lengths to standardize shipment processes and attempt to control variability in freight shipments.

Military and law enforcement units place a high value on standardized policies and procedures, as well as consistency in the employment of tactics, because they know from experience that this enhances security and predictability. The same holds true for transportation security, and entities engaged in transportation security work will proactively manage variability in an effort to enhance security. Monitoring and managing variability in the value chain is a core metric that can be used to create value and increase a firm's situational awareness of its assets and the reliability of its processes.

Velocity   The speed of throughput in the value chain.

Velocity in the value chain is arguably the single most important business metric for many transportation firms. In today's environment of just-in-time manufacturing, overnight delivery, and expedited shipping, speed of throughput can literally be money in the bank because carriers use speed of delivery as a primary measure of effectiveness and a sales tool, while shippers demand performance criteria related to timely delivery that can mean the difference between profit or loss on every shipment.

The VP of Distribution Operations at Target Corporation® emphasizes this notion that time is money, stating, "Target places significant business value on the smooth operation of our import value chain. If, at a port of entry, our containers are delayed by extended Customs review, there is a business cost to Target, and it could potentially impact our guests (customers). We manage that by having in place the necessary procedures and documentation to convince U.S. Customs that our container is secure, even if entering from a high risk country."[8]

The longer it takes for a shipment to move between origin and destination, the more opportunities there are to compromise that shipment. If one agrees that freight in transit is inherently vulnerable then it follows that speed through the value chain is also directly related to security. In other words, managing and monitoring the velocity of freight throughput represents an important metric related to value chain security because it defines the amount of time a shipment is at risk of being compromised.

Managing and monitoring the velocity of freight throughput represents an important metric related to value chain security because it defines the amount of time a shipment is at risk of being compromised.

Vulnerability   The level of exposure to value chain disruptions.

Security and vulnerability considerations are typically inseparable because one provides meaning to the other. Not only have transportation firms been forced to face an increasingly long list of threats (and related vulnerabilities), but the process of managing those vulnerabilities has proven to be increasingly important to an overall risk management strategy. At its most fundamental level, Total Security Management dictates that a firm that is better able to manage vulnerability can be considered a more valuable firm than one that is less able to do so.

Even before the modern terrorist threat, transportation firms had a requirement to monitor and manage value chain vulnerability in the course of managing their overall risk. Sadly, it seems that some paid little attention to this, and some paid none at all. Prior to 9/11 many transportation firms, including major carriers, did not even consider security to be important enough to warrant a full-time managerial position; and it was not unusual for value chain security to be relegated to a collateral duty for the person responsible for risk management. But even in those days there was a direct relationship between value chain disruptions (vulnerability) and business success. In today's environment, the linkage between vulnerability and business success has been elevated, and the importance of understanding and managing vulnerabilities is properly recognized as directly related to value within an organization.

The Four Security Ds

Security initiatives that counter threats to transportation assets can be organized into four primary areas, which are identified as the Four Security Ds:

Deter   The ability to discourage or prevent a value chain disruption.

Preventing value chain disruptions through security and business functions happens perhaps tens or hundreds of thousands of times every day in transportation firms throughout the globe. In fact, discouraging or preventing disruptions can, in some cases, be considered the core function of a person responsible for moving freight. Once a shipment has been coordinated, the main goal for most freight expediters is to discourage or prevent disruptions. Those disruptions, on the business side, are usually considered in terms of issues such as communication problems, missed pick-up and vessel sailings, routing issues, or lack of conveyances. All of these disruptions can affect levels of service and ultimately profitability.

The security manager is concerned with discouraging or preventing disruptions of another kind. Deterrence, in security language, has to do with taking actions that avoid a confrontation or event. Transportation security managers make a living by anticipating and then avoiding or discouraging potential threats related to theft, vandalism, sabotage, and terrorism. This can be achieved by a variety of basic security measures, including use of proper lighting and locks, random patrols, and proper inventory control measures. Such measures can discourage thieves, terrorists, and other criminals from disrupting the value chain—hardened targets are not attractive ones.

Detect   *The ability to discover the existence of a threat in the value chain.*

Good value chain managers must develop the ability to discover or anticipate threats to operations as early as possible. Interestingly enough, transportation firms are faced with plenty of threats on the business side of the house that are just as important to anticipate and detect as traditional security threats. Labor unrest, supplier discrepancies, natural disasters and weather issues, unscheduled maintenance requirements, and other traditional business threats should be routinely detected and eliminated or mitigated as a normal course of business.

Transportation security teams use essentially the same processes to detect security threats to the value chain. In either case, early discovery or anticipation can make the difference between a shipment that is delivered on time and intact and one that has been compromised.

Delay   *The ability to temporarily impede or hinder a threat to the value chain.*

Delaying a threat usually buys time to develop an effective response strategy. This is true in business matters as well as with security issues. Transportation managers develop strategies to temporarily impede or mitigate threats as a requisite for survival. Delaying a delivery cutoff can be the difference between making and missing a vessel sailing. Convincing a disgruntled truck driver to make one more run before quitting can save a shipment commitment. There are endless examples of how delay can be an important transportation business function.

By the same token, delay for a security professional can provide an opportunity to optimize the response to a transportation security threat. The ability to temporarily hinder or impede a threat to the value chain could be absolutely critical in terms of maintaining a transportation firm's reputation for safe and secure service. Employees responsible for operations or security functions have significant interest in employing the tactic of delay.

Dispatch   *The response to a value chain threat.*

Threats demand rapid, appropriate responses. It's that simple. Threats posed to core business functions, as well as security threats to the value chain, must be dealt with in an efficient and effective manner. Organizations that are better prepared to dispatch an appropriate response to a value chain threat are more likely to avoid or mitigate losses. In addition, proper response today often creates value by reducing future threats from terrorists and criminals, both of whom conduct preoperational surveillance and tend to select lesser-defended targets and ones that can't respond well to adverse events. Because firms that are better able to avoid or mitigate losses are more likely to avoid security events altogether or to properly deal with those that do occur, they should be recognized as more valuable than those that are not.

> **Dispatching an appropriate response to a threat creates value by reducing future threats from terrorists and criminals, both of whom conduct preoperational surveillance and tend to select lesser-defended targets and ones that can't respond well to adverse events.**

The Four Solution Set Cs

Finally, the TSM solution set can be used to formulate a strategy for creating value through more effective security initiatives, which are called the four Solution Set Cs:

Coordination   *Harmonious functioning of business and security initiatives for effective value chain results.*

Coordination is essential to all businesses, including transportation firms. Forwarders coordinate with carriers; intermodal managers coordinate with trucking companies; booking agents coordinate with loadmasters; and customer service representatives coordinate with all of them. In transportation, coordination is a core function that ensures success. The ability to coordinate effectively is an important business metric for determining a firm's ability to sustain operations, which in turn creates value in terms of brand equity and the ability to capture market share from competitors who are unprepared. It is no surprise then to learn that coordination is considered to be a "mission critical" element in the security world. Transportation security solutions can include an incredibly complex mix of jurisdictional, statutory, regulatory, and law enforcement issues. Firms that have mastered security coordination functions can, by default, expect to be better prepared to handle any security threat. Being better able to handle such threats translates to being a more trusted service provider, which in turn creates value for the firm in terms of brand equity.

Cooperation *Support given to and received from others for mutual security and business benefit.*

The ability to support others for mutual benefit is essential throughout transportation, which sometimes seems to be the last industry where "my word is my bond" remains a valued consideration. When good transportation professionals commit to a deadline or to provide a service, one can typically count on that commitment being carried out. As a result, many in this industry learn to trust and rely upon others in an effort to optimize their performance. Cooperation is a critical component of these freight operations.

Cooperation in security matters is no less important. The FBI shares crime data with railroad security managers to help each party better do their job and increase overall security. Port security managers have a requirement to cooperate with Customs and Border Protection [CBP] officers, agricultural inspectors, and local police on a daily basis. And truck driver participation in the Highway Watch Program is perhaps the ultimate example of how individual cooperation with the government can potentially result in transportation security benefits for many. Cooperation is an important component in value creation.

Consultation   *Seeking knowledge to improve business and security posture.*

Seeking knowledge to improve business processes and practices is another fundamental component of value creation for transportation. Individuals and companies routinely consult with one another in an effort to learn from and improve upon their performance. Industry working groups and professional conferences and trade shows are a mainstay in transportation and the primary conduit for the exchange of professional information.

Not unlike their counterparts on the operations or administrative side of a transportation business, security professionals also routinely use consultation as a tool to enhance security. Firms that have formal, systematic ways of improving security posture by consulting with others to derive best practices and compare policies and procedures to conventional wisdom are far more likely to benefit from the available body of knowledge in this field.

Collaboration   *Deliberately combining forces with others to improve business and mitigate risk.*

Last but certainly not least, collaboration is a business tool that lies at the heart of value creation for both business and security initiatives. Steamship lines are members of strong alliances that control freight markets and trade lanes. They are even provided with the unusual benefit of being exempted from anti-trust laws that can serve to inhibit collaboration in other industries. Railroads also employ formal and informal strategic alliances in order to optimize service and leverage resources. **To collaborate is to mitigate;** Collaboration is a hallmark of effi- **to mitigate is to create value.** cient and effective service in the transportation business.

By maintaining close business relationships and forming strategic alliances related to operational requirements, transportation firms also set the stage, knowingly or unknowingly, for employment of a powerful tool in the effort to mitigate and manage risk. Collaborative planning tools and practices, when properly employed, can provide a transportation entity with a powerful resource for combating the threats that attempt to disrupt or destroy their services. To collaborate is to mitigate; to mitigate is to create value.

Conclusion

TSM approaches the concerns of the modern age through a broad-based, holistic approach. Transportation and security professionals can draw on the concepts presented in this chapter by applying them to their own firm's individual circumstances, beginning with an assessment of the strategic and operational considerations and then moving on to the development of a more tailored tactical plan. In this way they can ensure the systematic use of the TSM framework in their approach to addressing resiliency and continuity challenges.

The Total Security Management approach requires a serious commitment from the entire business, but especially the management team and particularly during the early stages of implementation. In many organizations, individuals in a single department—be it security, logistics, sales, marketing, or any other—will not have the breadth of focus to perceive the full scope of the potential for value creation from various cross-departmental process improvements. Accordingly, TSM will require active dedication to the process of change implementation. In part, this is simply a reflection of the realities of organizational dynamics and a corollary of Newton's first law of motion. For, just as objects at rest tend to stay at rest, change is always difficult because many of the actors in any organization are more content to do tomorrow what they did yesterday than expend the effort to adjust to the dynamic realities of the ever-changing marketplace.

On the other hand, people with operational responsibility are often the ones who will recognize processes that are ripe for review, either because they are inefficient and redundant, or because they create undue risk. Security is a field where employee practices play an unusually prevalent role in proper implementation, and this inevitably enhances the relative importance of the role of the individual in TSM. Operators are also the ones who usually witness the benefits, firsthand, of improved business process and security initiatives.

For example, at security's most basic level, any lock is useless against thieves if employees don't properly close the door, and even armed guards can't protect lives unless checkpoints or roving security teams properly control traffic flow and screen delivery trucks to search for bombs and other threats.

As a result, it often falls to management to integrate and coordinate the competing priorities of the corporation into one cohesive strategy, including the intersection of business processes and security. In addition, senior leadership plays a major role in absorbing the costs of self-analysis and initial remediation implementation because they are in a position to understand that certain direct, immediate costs will be more than compensated for by gains over time in less tangible areas such as investor confidence and relatively higher credit ratings.

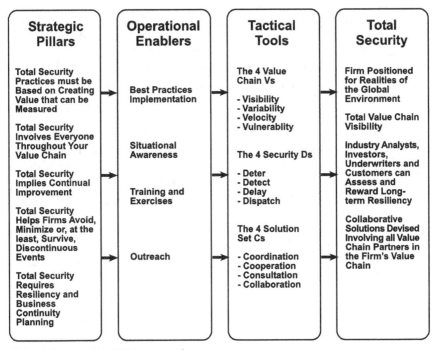

Strategic Pillars	Operational Enablers	Tactical Tools	Total Security
Total Security Practices must be Based on Creating Value that can be Measured	Best Practices Implementation	The 4 Value Chain Vs - Visibility - Variability - Velocity - Vulnerablity	Firm Positioned for Realities of the Global Environment Total Value Chain Visibility
Total Security Involves Everyone Throughout Your Value Chain	Situational Awareness	The 4 Security Ds	Industry Analysts, Investors, Underwriters and Customers can
Total Security Implies Continual Improvement	Training and Exercises	- Deter - Detect - Delay - Dispatch	Assess and Reward Long-term Resiliency
Total Security Helps Firms Avoid, Minimize or, at the least, Survive, Discontinuous Events	Outreach	The 4 Solution Set Cs - Coordination - Cooperation - Consultation - Collaboration	Collaborative Solutions Devised Involving all Value Chain Partners in the Firm's Value Chain
Total Security Requires Resiliency and Business Continuity Planning			

FIGURE 2.5   The TSM approach

With the backdrop of globalization, the three change agents, and the addition of the specifics of TSM's five strategic pillars, four operational enablers and the value creation model, it is time to take a closer look at the business case that underlies the Total Security Management approach (see Figure 2.5), including the specifics of how TSM can be used as a tool for value creation, rather than merely being a cost to the implementing firm.

Notes

1. George A. Zsidisin, Ph.D. and Gary L. Ragatz, Ph.D. and Steven A. Melnyk, Ph.D., "Effective Practices in Business Continuity Planning for Purchasing and Supply Management," 21 July 2003, <http://www.bus.msu.edu/msc/documents/AT&T%20full%20paper.pdf > (8 April 2006).
2. Russ White, "Study Finds Supply Chain Often Neglected in Planning," MSU News on the Web, 2004, <http://newsbulletin.msu.edu/supplychain.html > (8 April 2006).
3. LCP Consulting and the Centre for Logistics and Supply Management, Cranfield School of Management, *Understanding Supply Chain Risk: A Self-Assessment Workbook*, 2003, <http://www.som.cranfield.ac.uk/som/research/centres/lscm/downloads/60599WOR.PDF > (8 April 2006).
4. Deloitte Research, *Prospering in a Secure Economy*, 2004, <http://www.deloitte.com/dtt/cda/doc/content/DTT_DR_ProsSecFull_Sept2004.pdf#search='Deloitte%20Research%20prospering%20in%20the%20secure%20economy > (1 January 2006).
5. Yossi Sheffi, *The Resilient Enterprise* (Massachusetts: The MIT Press, 2005), 27–28.
6. Deloitte Research, *Prospering in a Secure Economy*, 2004, <http://www.deloitte.com/dtt/cda/doc/content/DTT_DR_ProsSecFull_Sept2004.pdf#search='Deloitte%20Research%20prospering%20in%20the%20secure%20economy > (1 January 2006).
7. Yossi Sheffi, *The Resilient Enterprise* (Massachusetts: The MIT Press, 2005), 106.
8. Diane Closs, "Security Practices Safeguard the Supply Chain," Michigan State University BROAD Business on the Web, 2005, <http://bus.msu.edu/alumni/publications/broadbusiness/05/reality2.cfm > (23 March 2006).

"Therefore, we strongly urge companies not to consider security investments as a financial burden, but rather as investments that can have business justification, that can result in operational improvements, and that ultimately may promote cost reduction, higher revenue and growth."

—Barchi Peleg-Gillai, Gauri Bhat and
Lesley Sept, *Stanford University*

Chapter Three

Creating Value: The Case for TSM

The maritime, trucking, aviation, and railroad freight transportation networks that crisscross the globe are critical segments of the world's shared infrastructure because of their significant role in sustaining the viability of the international economy. Transportation-related activity in the United States alone accounts for approximately ten million jobs and some eleven percent of the U.S. gross national product.[1] Global transportation networks are almost all controlled by private sector firms, however, which means that any significant changes must be viewed in terms of business imperatives and value creation. Because this fact has been largely ignored, it is not surprising that security threats to the global transportation networks have not been fully addressed. There are three primary reasons for this reality:

· Corporations have not made security initiatives a priority, and are not managing security as a core business function.
· Financial analysts, shareholders, and insurance underwriters have not defined a methodology for evaluating the relative security of a transportation enterprise, and for rewarding corporate security investments that create value.
· Transportation industry executives and professionals lack the strategic framework required to make informed security decisions that mitigate risk, and at the same time create value for their organization.

FIGURE 3.1   The top three impediments to securing transportation networks.

The Total Security Management approach has been developed as an enterprise-wide methodology that represents a preliminary solution to all three of these challenges. Specifically, it ensures that security practices incorporate fundamental management tools such as standards, benchmarking, best practices, security performance metrics, and fundamental business analysis. As such, TSM is a tool that the transportation industry can use to initiate the process of integrating security into all business units across an organization.

Indeed, there are early indications emerging that show that the major firms in the global marketplace are already beginning to modify their behavior and business practices and turn in this direction. In a report funded by AT&T, three professors at Michigan State University's Eli Broad College of Business determined that, "...the best organizations go beyond the use of tools and incorporate supply continuity planning and management as a philosophy—a way of thinking about supply chain management. They believe that it is through the proactive management of supply risk management, of which supply continuity planning is a critical activity, that firms can "bring light" to the dark side of managing supply chains. Supply continuity planning and evaluating security practices enable management to make effective and efficient supply chain management a part of everyday business, no matter what might happen on any given day."[2]

TSM follows directly along this argument, putting the philosophy into action by delineating both the actions and areas of the concerns that firms should consider in preparing themselves for success in the face of disruptive events. By embracing the five pillars of TSM, and using them to plan, implement, measure, and refine security initiatives, the transportation industry can manage risk while also creating value. Fundamental organizational change, guided by aggressive application of these TSM concepts, is the best way to transform transportation security from a drain on resources into a powerful tool for value creation.

> Fundamental organizational change, guided by aggressive application of TSM concepts, is the best way to transform transportation security from a drain on resources into a powerful tool for value creation.

As testament to the deregulation and privatization of many state or government controlled utilities during recent decades, it is generally accepted that governments are unable to control or sometimes even effectively regulate industries with enough agility to achieve their ends in the most efficient manner possible. Yet efficiency in all things is an increasingly important prerequisite for maintaining practices that are able to succeed in today's hyper-competitive global environment.

At the same time, those who own, control, and operate critical transportation infrastructure assets can no longer simply expect to operate with a "business as usual" mentality toward security. We now operate businesses in an environment where infrastructure assets such as waterways, border crossings, bridges, tunnels, and communications hubs are necessary to ensure continuity of business operations—and are also at the top of the target list for those who desire to disrupt the global economic system. Just-in-time processes and the increasingly

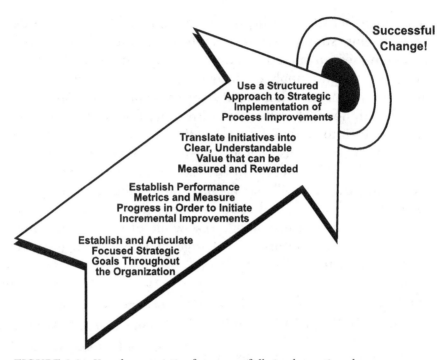

Successful
Change!

Use a Structured
Approach to Strategic
Implementation of
Process Improvements

Translate Initiatives into
Clear, Understandable
Value that can be
Measured and Rewarded

Establish Performance
Metrics and Measure
Progress in Order to Initiate
Incremental Improvements

Establish and Articulate
Focused Strategic
Goals Throughout
the Organization

FIGURE 3.2   Key characteristics for successfully implementing change.

complex web of global production and related processes for transporting goods have put an end to the time when hoping for good luck would suffice. As a result, creating value through security must be a deliberate, continuous effort focused on the desired end state.

The translation of this focus on value-added security is embodied by the process improvements required to implement TSM. This methodology empowers public and private sector executives to adopt a universal approach to security that proactively and effectively addresses transportation network and freight security issues. TSM also provides a framework for developing metrics that will serve as a foundation for establishing and evaluating transportation security initiatives. TSM provides insurance and financial analysts with tools to begin differentiating and rewarding firms that are effectively and efficiently managing transportation security requirements. And finally, Total Security Management practices provide context to the threats and the corresponding solutions that have the potential to enhance efficiency while simultaneously decreasing vulnerability.

TSM and Return on Investment

All expenditures in a firm are inherently tied to the need to provide a reasonable return on investment (ROI) for each dollar or other resource expended. However, that valued return on investment doesn't always need to manifest itself in the form of direct revenue, and can also include such significant factors as insurance against discontinuous events, increasing brand equity, or better long term market share.

Traditionally the business case for justifying investment in various security measures has relied on two factors: presumed savings in the event of a catastrophic discontinuous event and the reduction of theft and petty crime. These remain significant factors in the determination of appropriate security investment levels, for the costs of being ill-prepared for economic or value chain shocks can be dramatic, even terminal, and no business should willingly accept high levels of theft that literally drain the corporate coffers. Unfortunately, decisions to invest in day-to-day security and catastrophic preparedness often lose out to expenditures on advertising, research, or other activities that are presumed to create more immediate and assured economic rewards.

According to Deloitte Research, however corporate investments in secure commerce can go hand-in-hand with the following measurable business benefits:

· Cost reduction, as driven by security-related investments that increase efficiency, such as IT tracking tools used to streamline shipping costs.
· Enhanced revenues, as driven by monitoring technologies that reduce shortages by making inventory awareness real time and highly accurate.
· Better risk management, as driven by proactive security policies.

- Brand protection, as driven by investments in incident prevention and crisis response.
- Market share preservation, as driven by faithful adherence to best practices and regulations regarding increased security.[3]

The TSM approach, in recognition of the need to offer more measurable and concrete benefits from better security practices implementation, focuses on creating near-term ROI from TSM-related activities whenever possible. This is significant because security-related expenditures generally enable, but do not directly increase, a firm's profitable operation. Although the logic behind the TSM process intuitively indicates that holistic security and resiliency of an enterprise should produce measurable market gains, at present proving such gains is arguably a difficult and non-standard process.

Part of the problem arises because it's difficult to develop concrete guidance for determining which TSM-related activities will offset potential risk, how much risk will be offset, and at what cost. But we know that market analysts and others who influence value analysis such as credit ratings rely upon specific risk ratios and calculations to drive their evaluations. The challenge exists in associating TSM inititives to measurable value metrics. Until and unless the longer-term benefits of TSM initiatives can be measured and can be demonstrated conclusively through economic gains, TSM-related expenditures will need to create enough value to at least partially offset their costs. Nelson Repenning and John Sterman entitled their study of improving operational processes "Nobody Ever Gets Credit for Fixing Problems that Never Happened...".[4] This phrase sums up how investment in security has often been viewed, but under the TSM approach value *can* and *should* be recognized for it's ability to prevent disruptions and mitigate against potential impacts.

TSM offers a systematic framework for process improvements and a more reasoned selection of trade partners based on their relative reliability in the face of adverse events, including business continuity strategies. It also supports a closer collaboration with value chain partners, and their partners, which can result in additional benefits. These benefits can include building confidence among key firms as well as creating efficiencies gained from repeated business among the same vendors and customers. TSM also directly impacts the critical external shareholders that affect a firm's ability to raise capital.

Rewarding TSM Practices: Financial Analysts, Insurers, and Regulators

Implementing TSM practices should make a firm a more attractive trading partner to its value chain associates, but the realm of interested stakeholders goes far beyond industry partners. This fact opens the door to another major way in which the TSM approach can positively influence a firm's ROI. As a holistic framework for evaluating value chain resiliency, TSM provides a starting point to begin to directly measure created value—a starting point that then enables in-depth analysis by key external stakeholders, including financial analysts/investors, insurance underwriters, and regulators.

Wall Street analysts are critical because they develop the evaluations that guide the investors and drive credit ratings and investment opportunities. The analysts' primary function is to compare endless columns of data on various companies, measure them against market performance, and then determine whether their firm, and in turn various individual investors, should or should not invest their money in a given business.

Who are these critical players and how do they determine a firm's worthiness? The U.S. Department of Labor describes financial analysts as follows:

> Financial analysts assess the economic performance of companies and industries for firms and institutions with money to invest.... [they] work for banks, insurance companies, mutual and pension funds, securities firms, and other businesses, helping these companies or their clients make investment decisions. Financial analysts read company financial statements and analyze commodity prices, sales, costs, expenses, and tax rates in order to determine a company's value and to project its future earnings. They often meet with

company officials to gain a better insight into the firm's prospects and to determine its managerial effectiveness. Usually, financial analysts study an entire industry, assessing current trends in business practices, products, and industry competition. They must keep abreast of new regulations or policies that may affect the industry, as well as monitor the economy to determine its effect on earnings.[5]

The description goes on to say that "Mathematical, computer, analytical, and problem-solving skills are essential qualifications." While all this is true, what this description doesn't mention is that financial analysts are often overworked and face extremely tight deadlines. Nor does it mention that, although these analysts tend to be on average very bright, they are not often specifically trained or directly experienced in the field they analyze, such as transportation or security. Many analysts could be described as industry generalists that specialize in research and statistical modeling. As a result, they make predictions relying on data and trends that use the past to extrapolate what the future will look like, by evaluating variables such as cost per dollar of revenue and earnings per share. But they don't always develop a means to reward those firms that are taking on additional short-term burdens to be better prepared for a risky future. As the first few chapters have pointed out, it is essential for firms to prepare for the medium and long-term effects of an ever more fragile economic system that relies, first and foremost, on the relatively timely delivery of goods.

The role of the insurance industry should not be underestimated either. Its power to dole out economic rewards and penalties can be a major factor in the viability and profitability of any firm. This is primarily due to the fact that all sizeable public companies have to answer to other interested parties, notably these same analysts, investors, and insurance companies, in order to raise capital to fund growth. Typically, insurance companies apply complex formulas based on the predicted *frequency* and *severity* of events in order to determine appropriate average risk over time. In the case of significant discontinuous events that affect the global transportation network, however, these two factors are nearly impossible to predict with any accuracy. Consequently, insurance providers have a tough time insuring against such events and the potential residual trail of directly and indirectly related costs to their clients.

The members of the various regulatory bodies who have control over international trade and the flow of goods make up the third major group of external stakeholders that can derive value through the application of TSM principles and practices. TSM provides a framework for documented and verifiable security processes that demonstrate to regulatory bodies that if a firm adheres to the best practices for transportation security, the firm is better able to establish that its enterprise is secure. At the same time, the regulatory body can make the determination that the firm is adequately managing risk. In addition, TSM facilitates appropriate outreach activities that support interaction with these various regulatory partners who are, after all, an embedded part of the value chain.

For example, reduced inspections associated with the U.S. Department of Homeland Security's Customs-Trade Partnership Against Terrorism (C-TPAT) have become a significant cost-driven (i.e., value-creating) source of ROI for several firms. Hasbro, the toy manufacturer, spent just under $200,000 to achieve its upfront C-TPAT compliance status and spends an estimated additional $112,500 a year maintaining it. However, since it became C-TPAT-certified in November 2002, its inspections have dropped from 7.6 percent of containers coming into the U.S. in 2001 to 0.66 percent in 2003. Given that in 2003 the company imported about 8,000 containers, and that inspections can cost the firm on average $1,000 per occurrance, Hasbro is saving almost $550,000 a year in inspection costs alone, a nearly 5-to-1 return rate.[6] This kind of process improvement is precisely the kind recognized and supported by the TSM approach.

Additional value creation can be seen if the definition of "value" is expanded beyond the idea of dollars and cents. Below are several of the major areas in which TSM investments will create value by supporting better processes that enrich the firm in a variety of ways.

Catastrophic Preparedness

It can be difficult to demonstrate the tactical creation of value related to benefits derived from expenditures aimed at resiliency. These inititatives are undertaken to avoid or mitigate major discontinuous events that can not be accurately predicted. In this case, demonstrating value requires an attempt at "proving the negative"— or placing a value on a savings that has not yet occurred. However, it remains paramount that firms invest in catastrophic preparedness measures. This is a function of both the threat and the changing nature of the global system as embodied in the three change agents described earlier in this book. As noted by Stanford's Lee and Whang, "If ports and border crossings were closed for a meaningful time after a major terrorist attack, the economic impact would be devastating. It is not possible to quantify the full direct costs of damages and casualties, recovery measures, congestion, and disruption to business and daily life."[7] The only way to survive such events and be prepared for the immediate resumption of trade is to be able to demonstrate compliance with best practices for security processes. Significantly, trying to implement such changes from scratch at the time of the next significant discontinuous event will require time and resources that a firm may not have available.[8] Accordingly, TSM requires that firms address their value chain processes in advance of future events, as well as to develop appropriate business continuity plans.

Another important but hard to measure benefit of thorough disaster preparedness is the potential for a business to gain the edge against its competitors by better positioning itself for resumption of operations following a disruption. In a world where minutes matter, building redundancy into the firm's communications system or value chain can create great

Don't Just Survive - Thrive!

TSM also creates value by enabling firms to make profits, capture market share, or even expand into new markets in the wake of a disruptive event. Following any such event there are those firms who come out ahead and those who lose; each firm's goal is to make sure it is the one on top. One of the major reasons TSM investments make sense now, in this time of increasing globalization and interconnectedness, but in advance of the next catastrophic event, is that being ready, responding well, and recovering quickly will enable a firm to capture market space from other firms who failed to make similar preparations. This is the real-world embodiment of what has been called "the three little pigs effect," meaning that no one should begrudge the gains of the firm who had the foresight to invest properly in security and resiliency, and was therefore better positioned than those who did not bother to build their houses out of brick.

market power because firms that respond well to crises, and are able to continue business, often gain market share from their competitors—and once customers switch to the new provider they are unlikely to switch back.

Theft Reduction and Process Improvements

Today, savings from reduced product losses (shrinkage) can compensate in part for the cost of implementing various security measures. However, TSM calls for analyzing and improving security processes such as real time and comprehensive value chain visibility and tracking, which can in turn enable even greater reductions of the shrinkage of goods. Firms may no longer need to build shrinkage into the operating budget, because they will have the ability to account for and track goods, at the item level. Unlike terrorism or certain other discontinuous events, shrinkage is a continuous event in transportation and can be directly managed. Effectively controlling theft-related process deficiencies—much the same way that defects in a manufacturing process are eliminated using TQM processes—should minimize the impact of such events, creating value over time through decreased losses.

The openness and benefits of better visibility into the firm's value chain operations will also open up opportunities in unexpected ways, especially through collaborative efforts with partners. For example, in 2002, sixteen Scottish whisky distillers who were all facing similar low-level security problems banded together to hire a single firm that collected each of their loss data for review. The consultants determined that all sixteen distillers were facing similar losses along the same delivery route, but that they never discovered it because individually the losses were below each firm's threshold for further review. As a result of this collaborative security approach, the firms in question found the anomaly and put an end to their losses.[9]

Reaching out to the law enforcement community also can help firms to address the low-level nuisance crimes such as theft and counterfeit goods that can have a significant economic impact over time. For example, Target's Director of Supply Chain Assets Protection reports that "[At Target] we embrace opportunities to get involved in law enforcement partnerships that impact organized criminal groups, even when those groups are not directly impacting our business...at least not yet. There's simply no competitive advantage to allowing organized criminal elements to flourish."[10] This kind of cross-industry teamwork helps make global businesses more efficient all around, which inures to everyone's benefit over time through aggregate cost-savings.

Fewer Inspections

TQM creates ROI by steering manufacturing firms away from the traditional but inefficient practice of

> TQM creates ROI by steering manufacturing firms away from the traditional but inefficient security practice of inspecting for deficiencies using random inspections. Likewise, TSM builds in security process improvements that ensure more secure end-to-end processes throughout the entire value chain, mitigating the need for spot checking for deficiencies.

checking for deficiencies using random inspections. According to Lee and Whang, "The quality movement started with the recognition by industry that defects can be very costly to a company. The costs of product failures out in the field, or 'external failure costs,' that include customer process downtime, catastrophic impacts to society, liabilities, product recalls, field repair, brand equity damages, and adverse effects to future sales, can be far greater than the product cost itself."[11] Professor Yossi Sheffi of MIT echoes this recognition, stating:

> "Focus on security measures can enhance the efficiency of the supply chain in the post-9/11 world just as the focus on quality enhanced efficiency. The quality movement has focused on building quality into the product rather than trying to inspect for defects later; it stressed process integrity so that if there is a problem on the production line, the line stops and the problem is corrected before multiple defective products would roll off the assembly line... The security lessons are obvious: fighting problems at the source (for example, securing freight origins and departure ports); continuous monitoring for anomalies (for example, in shipping patterns and people's behavior); and the development of security culture throughout the organization."[12]

Much as TQM encouraged firms to determine better processes for production, which in turn put out better products, and reduced the need for costly and inefficient quality spot-checks, TSM builds in security process improvements that ensure more secure end-to-end processes throughout the entire value chain. This is a significant improvement for the long term because it rectifies the root of the problem and eliminates the source of the vulnerability, as opposed to merely looking for symptoms of the security process failures by randomly spot-checking for evidence of deficiencies.

Ready for the Worst, Ready for the Routine

One of the most important, though often overlooked, ways in which TSM creates value is that by ensuring an ability to handle major disruptions and crises it inherently makes dealing with a panoply of more common issues increasingly more manageable. In other words, if firms are ready for the worst, the routine becomes just a matter of course, and in fact becomes a good way to test and exercise their readiness.

TSM creates value by instituting the appropriately flexible and resilient management practices that support effective crisis operations and also enhances an organization's capabilities regarding day-to-day operations.

Furthermore, concentrating security and readiness improvement initiatives on the key elements in transportation that drive the business—and subsequently need to be protected, such as *fixed assets, assets in transit*; *brand equity*; and *human capital*—inevitably leads to improvements in daily operations. By focusing on these core business elements, firms can also identify additional opportunities for improvement, including the ability to take a broad view of corporate-wide present and future market positions. In this manner, TSM and related business practices create value because instituting the appropriately flexible and resilient management practices that support effective crisis operations also enhances an organization's capabilities regarding day-to-day operations.

Good Security Promotes Good Safety

Towards the end of the U.S. industrial revolution, it became conventional wisdom that the costs of implementing appropriate safety processes could be more than offset by the savings gained from reduced liability payments, employee turnover, and missed workdays. This is significant because the tenets of TSM are in part cost-justified because they follow the old security adage that "good security is good safety."

Just as was the case when they began to focus on specific production quality, firms will now find that the incremental advantages and the momentum of implementing good security, sound resiliency planning, and reasonable disaster preparedness exercises will create momentum in the direction of proper safety as well, including carrying out routine maintenance and related activities on the same regular schedule as other preparedness activities. So many of the same processes and checks that encourage proper safety and good maintenance lend themselves quite well to the processes and activities that define appropriate security measures. This association can serve to blur the line differentiating safety from security or perhaps render it non-existent.

Part of the justification for this correlation between security and safety can be thought of as a derivative of James Q. Wilson and George L. Kelling's famous Broken Windows Theory, which postulates a "slippery-slope" theory of crime. Specifically, the theory contends that, in a neighborhood where a building is left with a single broken window that remains neglected, there is a much greater likelihood that other windows also will be broken in short order; conversely, if care is shown and the broken window is repaired, and the trash is picked up, and other minor offences are properly regulated, then the larger crimes tend not to appear there. They believe the same holds true for other issues, such as graffiti and loitering, because they are all signs to the community that no one cares, which in turn creates an opening for more serious and violent crimes.[13] In a similar manner, the adherence to proper security procedures reinforces good safety, and vice versa.

Given that all transportation firms have a variety of international and other specific safety codes they must follow

Key Characteristics for Successfully Implementing Change

- Use a structured approach to strategic implementation of process improvements
- Establish performance metrics and measure progress in order to initiate incremental improvements
- Establish and articulate focused strategic goals throughout the organization
- Translate initiatives into clear, understandable values that can be measured and rewarded.

(including OSHA regulations), the fact that TSM practices reinforce and even help justify further investment in safety measures enables firms to gain both in security and safety from their TSM initiatives.

The TSM Problem Solving Process

Every TSM initiative, if properly planned and executed, should use the problem solving process outlined below as a guideline. This sequence is a continual loop with eight key steps:

1. Identify and evaluate security imperatives for the firm
2. Perform security gap analysis to identify focus points for ROI
3. Create measurable security metrics that apply to these focus points
4. Implement security initiatives that drive the needed changes
5. Evaluate changes and measure progress
6. Incorporate successful initiatives into long-term continuity plans
7. Share information about successful initiatives with value chain partners
8. Begin the process over from step one

This cyclical process of continual improvement enables managers to evaluate progress and identify real return on investment.

Transformation of transportation security can occur by leveraging TSM processes in order to improve and enhance existing capabilities to combat value chain threats. Public and private sector executives and managers can use this model to develop programs that can harness the latent power that exists throughout an organization, applying it wherever business process improvement and security concerns intersect. By applying the core elements of TSM, organizations can optimize their investment in security resources in a way that maximizes their impact, and ensures a standard, coordinated approach to transportation security. Ultimately, by taking actions within an

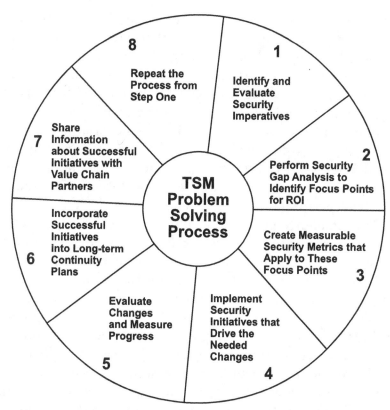

FIGURE 3.3   TSM Problem solving process.

organization that eliminate or minimize disruptions, enhance resilience throughout the value chain, and enable business process improvement in the context of security, organizations will be able to claim increased value, and should be rewarded accordingly by those with a vested interest in the corporation.

Conclusion

TSM practices are guided by the overriding need to create value for the firm. They do so through a variety of means, including traditional security functions such as increased preparedness for significant discontinuous events, enforcement of disciplined access control policies and procedures, and routine reduction of theft and petty crime. TSM also enhances value in less direct ways. Many businesses include the sale of products and services in their core functions, which requires building and maintaining positive relationships with both vendors and customers. These vendors clearly have an interest in establishing strong relationships with firms that will be long-term partners because the transaction processes of finding new partners and continually negotiating new deals can represent a drag on economic efficiency. Similarly, most consumers prefer to establish some degree of brand loyalty, if only to provide consistency in their expectations of a product's price and performance.

This challenge of assuring value chain partners and others of a firm's resilience and preparedness also can be seen as an opportunity. Multiple partners, service-providers and end users all have some vested interest in the firm's well being. This fact creates a bond, a mutual dependency that if approached properly can drive the various parties together, as opposed to driving them apart. For example, it becomes an opportunity for success for a transportation firm to engage a major distributor, develop common expectations for emergency response scenarios, and agree to collaborate if a disruptive event should take place. These activities can serve to strengthen the relationship between the firms, while at the same time increasing the likelihood that both parties can mitigate the impact of a discontinuous event. Indeed, the fact that at least one major client decides to stay with the affected firm will often entice others to do the same. The multiplier effect that can be generated by having other business partners committed to supporting the recovery process can go a long way towards buying the time needed for the resumption of normal operations. This hedge on resiliency can also be used to differentiate the enterprise.

Although it is essentially self-evident that many parties—including vendors, customers, analysts/investors, insurers, and regulators—have strong mutual interest in gauging the security and survivability of the enterprise, it may be somewhat surprising to learn how little time and effort is spent in pursuit of these activities. This is a result of several factors, but perhaps two above all others:

- In the ever more fluid market, investment analysts and others make determinations about corporate strength that are driven by immediate, short-term performance, as opposed to taking the long view of earnings and stability.
- There is no common, easy, and verifiable mathematical formula by which to evaluate various firms for preparedness and survivability for discontinuous events.

These two factors can represent a significant roadblock in the adequate assessment of the relative value of a firm's resiliency and preparedness for disruptive events. Any solutions that eliminate these problems should, by default, enable opportunities for critical external stakeholders to more effectively and comprehensively analyze a firm's security and resiliency.

The Total Security Management approach was devised to fill this need. TSM provides a customizable framework for systematic, deliberate process improvement initiatives that encourage standardized, verifiable best practices through the specific lens of the five pillars and the four tactical enablers. Then, to round out the business case, the TSM Value Creation Model delineates the specific value chain, security, and solution set metrics that define improvements to transportation security. In so doing it provides the means by which analysts, stakeholders, regulators, and underwriters can better approach the issue of evaluating and placing a premium on value chain resilience and preparedness for significant discontinuous events. This has been a significant missing link in the justification for security becoming integrated across the enterprise as a core business function.

> **TSM provides the means by which analysts, stakeholders, regulators, and underwriters can better approach the issue of evaluating and placing a premium on value chain resilience and preparedness for significant discontinuous events.**

Notes

1. David S. Lawyer, "Automobile Costs and Subsidy (A critique of Stanley Hart's writing," 1984, <http://www.lafn.org/~dave/trans/econ/auto_subsidy_shart.html> (25 April 2006).

2. David Closs and Gary Ragatz and George A. Zsidisin, "Sustaining business through disruption," Michigan State University Eli Board college of Business on the Web, 2005, <http://www.bus.msu.edu/alumni/publications/broadbusiness/05/reality.cfm> (3 March 2006).
3. Deloitte Research, *Prospering in a Secure Economy*, 2004, <http://www.deloitte.com/dtt/cda/doc/content/DTT_DR_ProsSecFull_Sept2004.pdf#search='Deloitte%20Research%20prospering%20in%20the%20secure%20economy> (3 March 2006).
4. Nelson Repenning and John Sterman, "Nobody Ever Gets Credit for Fixing Problems that Never Happened," California Management Review, 43 no. 4 (2001), <http://web.mit.edu/nelsonr/www/Repenning=Sterman_CMR_su01_.pdf#search='Nobody%20Ever%20Gets%20Credit%20for%20Fixing%20Problems%20that%20Never%20Happened%20%3A'> (25 April 2006).
5. Bureau of Labor Statistics, U.S. Department of Labor, "Financial Analysts and Personal Financial Advisors," Occupational Outlook Handbook, 20 December 2005, <http://www.bls.gov/oco/ocos259.htm#nature> (22 April 2006).
6. Ben Worthen, "Customs Rattles the Supply Chain," CIO Magazine, <http://www.cio.com/archive/030106/supply_security.html?page=7> (12 April 2006).
7. Hau L. Lee and Seungjin Whang, *Higher Supply Chain Security with Lower Cost: Lessons from Total Quality Management*, October 2003, <https://gsbapps.stanford.edu/researchpapers/detail1.asp?Document_ID=2273> (4 March 2006).
8. Rick Van Arnam and Robert Shapiro, "Cargo Security - The Legal Perspective: A Top 10 List," Metropolitan Corporate Counsel 11 no. 8 (2003), <http://www.barnesrichardson.com/files/tbl_s47Details/FileUpload265/39/CTPAT-Top10List.pdf> (5 March 2006).
9. Yossi Sheffi, *The Resilient Enterprise* (Massachusetts: The MIT Press, 2005), 141.
10. Kelby Woodard, "A Strategy of Trust," Prevention Magazine, November-December 2004, <http://www.cargosecurity.com/ncsc/ncsc_dotnet/articals/Global%20_Supply_Chain.pdf> (13 April 2006).
11. Hau L. Lee and Seungjin Whang, *Higher Supply Chain Security with Lower Cost: Lessons from Total Quality Management*, October 2003, <https://gsbapps.stanford.edu/researchpapers/detail1.asp?Document_ID=2273> (1 March 2006).
12. Yossi Sheffi, *The Resilient Enterprise* (Massachusetts: The MIT Press, 2005), 132-133.
13. James Q. Wilson and George L. Kelling, "Broken Windows: The police and neighborhood safety," *The Atlantic Monthly* 249, no. 3 (1982), <http://www.theatlantic.com/doc/prem/198203/broken-windows> (24 April 2006).

"Risk comes from not knowing what you're doing."

—Warren Buffett, *financial investment expert*

"Even a correct decision is wrong when it was taken too late."

—Lee Iacocca, *former CEO of Chrysler*

Chapter Four

The Risk Management Approach to TSM

Unpredictable but highly disruptive events can occur at any point along a firm's value chain. Furthermore, due to globalization and the advent of international economic competition in the production and distribution of goods and even services, the potential losses from discontinuous events can be magnified by the potential for far-reaching cascading effects that can cause systemic, industry-wide network failures. This includes risks created by everything from weather that impedes the suppliers and transportation providers from delivering raw materials to manufacturers, to potential terrorist events that disrupt the finance and banking industry and the processing of point-of-sale credit card transactions. Because disruptive events can take place at any time or place and can affect any industry, the risks are omnipresent. Fortunately, there are security and resiliency solutions that firms can implement to better position themselves to survive such events in order to protect against such threats.

As the famous financial investment guru Warren Buffet has noted, "Risk comes from not knowing what you're doing."[1] This is why the most secure businesses approach the task of securing fixed assets through the *risk management process*, which is a critical tool in the TSM approach. The goal of risk management is to empower senior decision makers with a rational basis for determining which risk mitigation measures to enact versus how much and where to accept other risks. The heart

FIGURE 4.1   The risk management cycle

of the matter involves the trade off between up-front security costs versus the *potential*, but potentially catastrophic costs represented by a significant discontinuous event.

Risk management begins by identifying critical assets, assessing threats and vulnerabilities, and then devising strategies to spend what resources are available in the most efficient and effective manner. Risk management's track record of success is strong because it enables businesses to understand, evaluate, and make trade-offs among their various types of risks, and to deliberately decide between investments for maximum risk reduction in one area and the acceptance of risk in another. Indeed, the risk management approach was developed to examine both the most important *and* the weakest links in a chain, and to develop a comprehensive approach that improves overall enterprise-wide preparedness for disruptive events. As a result, the process is equally valid for all types of risks, whether they're man-made or natural. This is true because terrorist attacks, hurricanes, tornadoes, mistake-induced power outages, and other events all require certain basic continuity of operations efforts.

Defining and Assessing Risk

According to the Cranfield University School of Management in Great Britain there are five defining characteristics of today's global environment as they relate to value chains and which dictate that each firm must undertake its own specific vulnerability assessments:[2]

The five characteristics are as follows:

1. There are many types of risk in the end-to-end value chain.
2. The characteristics of each risk in terms of its probability and severity varies greatly.
3. Risk is sensitive to the context of the company, its markets, and its position in the value chain.
4. The permutations and combinations of risk are such that few generalizations apply.
5. Pinpointing all of the areas of risk that a company may face is likely to be a difficult task.

As a result of these factors, risk management for each firm requires a strong understanding of the nature of risk itself. The first course of business in discussing risk is to correct the common misconception that the term "risk" is merely another way of referring to the likelihood of an event occurring, when in fact that is only half the story. Risk is actually a function of two equally significant variables: the likelihood of an event taking place, *and the severity of the event if it does occur*. This understanding leads to the realization that one can minimize risk by reducing *either* (or both) the likelihood or the severity of an adverse event. As a result it is an equally valid risk management strategy, for example, for a firm to reduce the impact of a tornado that could destroy a data storage center by building tornado-resistant buildings or by using two geographically-separated data centers. In this case, the firm cannot reduce the likelihood of the event taking place (assuming that relocation to a less tornado prone geograpic area in not an option), but it is empowered to take steps to reduce the severity of the event.

The need for corporate action is built on more than just the threat of the disruptive event itself, however. Legal liability considerations also play a significant role. According to security analyst Kevin Coleman, terrorism has become a foreseeable risk, and this creates legal ramifications for companies. Specifically, he says, "It means that organizations will be held to a higher standard of care. That means that entities that fail

How Risk = Likelihood * Severity

Why does assessing risk require measuring both likelihood and severity? What if you were running late and 75 percent more likely to get a speeding ticket, but that ticket would only cost one dollar? Would you speed anyway? How about if you had only a 25 percent chance of getting a ticket, but that ticket would cost ten thousand dollars? The difference is that, just like likelihood, severity matters.

to prepare for a terrorist attack could be held liable for their negligence in the legal system. Planning, education, and compliance are three key areas that can help protect executives and organizations from having suits brought against them in the event of a terrorist attack."[3] TSM addresses these same issues because just as terrorism is now deemed foreseeable, other types of disruptive events also are simply too prevalent to plead that they were "unforeseeable."

One of the most useful tools for evaluating risk is called a *vulnerability assessment.* Many vulnerability assessment tools that are currently being employed have been modeled after a Department of Defense/Special Operations Forces methodology that was originally created by Sandia National Laboratories as a target selection tool for dismantling key components of an enemy's infrastructure, such as armaments manufacturing, electrical power production, or airfield operations. Over time this approach has been adapted into specific formats relevant to various users, including the chemical, petrochemical, oil, and water treatment industries, as well as by the Department of Homeland Security for use with stadiums and other public gathering places.

The composition of an assessment team can vary based on the size and complexity of the entity being assessed, and may include external security and critical infrastructure dependency experts. However, it is imperative that all teams include several internal personnel who are knowledgeable of the firms' processes and business domain, and are also empowered by senior management to elicit the necessary details for analysis of existing practices.

A vulnerability assessment team should begin by examining existing security Standard Operating Procedures (SOPs), current training programs and tactics, and security and operational techniques as well as critical infrastructure dependencies. The next phase of an effective assessment entails comprehensive data collection and operational efficiency reviews in order to enable optimal risk mitigation. The team should compare the firm's activities with the best practices of similar companies, identifying key efficiency inputs for relevant risk reduction models, and quantifying the risk impact of various activities. The final phase of any assessment should include analysis and identification of critical corporate activities to determine an optimal process for linking estimated risk, readiness, and resource levels to actual operational requirements.

Value Chain Risk Mapping

Value chain risk maps are used to identify the structure of the value chain in two ways: by listing all relevant partners, and by identifying critical chokepoints where challenges may exist in the value chain. These risk maps are important tools for looking beyond internal and first tier value chain partners in order to thoroughly examine the detailed extended chain of relevant associated entities. This way, firms can identify inefficiencies, allowing them to address redundant processes, opportunities for consolidation to achieve bulk-buying discounts, and logistics coordination improvements.[4] Risk mapping is a detailed and intense process, especially if the partners themselves haven't conducted such a process to identify their segment of the value chain risk map. However, a variety of specialized consultants or self-guided commercial software programs also can be used to facilitate value chain mapping.

Knowing one's suppliers, whether they're contractual business partners or third party providers, is another important component for ensuring the security of a firm's entire value chain. Companies should be certain that consideration is given to security capabilities when selecting suppliers. Selection is often limited, so knowledge of the actual level of security provided and the security practices of the supplier becomes important. Detailed knowledge of the suppliers' policies and procedures, as well as information about the implementation and execution of those procedures, can be invaluable in managing risk. Companies should establish some base requirements for selecting suppliers based on criteria such as personnel screening, licensing review, background checks, documentation of saftey and security programs, and credentialing and access control procedures among other things. One of the goals for establishing these relationships should be to create an environment of mutual collaboration that will foster the growth of better security practices for both entities over time.

Companies should be certain that consideration is given to security capabilities when selecting suppliers. Firms should consider specifying a minimum level of security in contracts as a condition of doing business. The goal should be to create an environment of mutual collaboration that will foster the growth of better security practices throughout the value chain over time.

Firms should consider specifying a minimum level of security in their contracts, as a condition of doing business. Businesses should also perform security assessments and site visits in order to verify the level of compliance and to cooperatively develop plans to improve security. Firms should also consider working with suppliers to provide education and training in order to raise the level of awareness throughout the supplier's enterprise and to enlist the support of each employee to fulfill their role in security assurance. If the company is working through a third party logistics provider, steps should be taken to ensure downstream compliance. This too can be part of the contractual relationship that is established with business partners. Measurable goals can be developed so that suppliers can perform self-assessments to fix areas that need improvements and to demonstrate their security worthiness relative to their competition. As an incentive, particularly in the early stages of the relationship, a firm can reward good security practices, or

Risk Factors for Value Chain Partners

- Does the firm have multiple sources for critical components?
- Will inventory levels support sales during a prolonged disruption of up to several weeks?
- Has the firm prearranged for a variety of suppliers and customers for key parts/products?
- Does the corporate culture support innovation in the value chain?
- Are middle managers empowered to report minor disruptions before they become large problems?
- Are response plans and data backup procedures tested?
- Has the firm identified work-around solutions in the event of fuel or other shortages?
- Does the firm actively benchmark themselves against best practices of their peer competitors?
- Does the firm regularly update its business continuity plans? Do these plans include provisions for family members?

even incremental improvements in security as measured by the metrics, with a rate premium or additional business volume, thereby recognizing and rewarding the value of security in its partners. This kind of activity should, in theory, encourage more suppliers to begin to offer higher levels of security, and over time, market forces can establish these initiatives as a business discriminator.

Measuring Likelihood

Predicting the likelihood of an adverse event taking place is a significant part of a vulnerability assessment. However, this is an imprecise science at best, and one that requires a complex analysis of multiple variables. Two significant components are involved in evaluating the likelihood of the occurrence of such events: an analysis of historical patterns and an analysis of reasonably plausible threats.

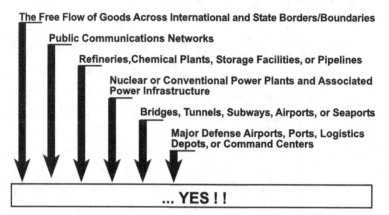

FIGURE 4.2   Risk Factors

In the intelligence community, historical patterns are often used as a guide for predictive analysis of future events. This is especially useful in terms of weather or naturally occurring events because they tend to result from cyclical patterns that can be traced over time. The firm must be aware, however, that its time horizon must be sufficiently broad to encompass long-term cyclical variations; for example it may not be the previous five years' hurricane seasons that are a fair predictor of the next season's, but rather the average intensity of storms over the past several decades that serves as the more accurate predictor of severe weather. In addition, the firm must be careful to adjust predictions based on recent changes to the factors influencing the historical record. For example, it might be considered a strong indication of proper safety practices if the historical record indicates that a certain critical highway through a distant transportation node is relatively safe because there have been no significant HAZMAT spills in the recent past. However, that data is not a reliable predictor of future events if a large new plant opens that requires trucking unprecedented amounts of HAZMAT through that same route. Accounting for such variables and making the assumptions as explicit as possible is a key factor in accurate forecasting based on historical patterns.

The calculations become even more complex when man-made factors, such as labor disputes and terrorist events, come into play. How can one predict the likelihood (or duration) of work stoppages at critical nodes on the value chain, as happened with the previously mentioned West Coast dockworkers lockout in 2002? Or consider the complexities of evaluating the risks of reliance upon an airline that has been previously targeted by terrorists. Is American Airlines at greater risk simply because of the propaganda value of attacking an airline with such a self-evidently American name, or is it a less likely target because terrorists already hijacked two of its planes on 9/11, and is therefore likely to be a relatively more vigilant airline? These are complex matters, and professional analysts and experts, including the U.S. intelligence community, have a hard time reaching a consensus.

This unpredictability factor comes into play in unexpected ways—as experienced by some of the financial industry firms that fled New York City after 9/11. One firm in particular had just recently moved their headquarters to the New York area prior to 9/11, and within a year had returned to the Midwest. While this was a wise calculation for overall risk reduction (New York City being openly acknowledged as a prime Al Qaeda target), that same firm may still be at risk because the level of security presence and awareness in their Midwestern location is so much less than that in New York. This might inadvertently make them a relatively easier target.

Consideration must also be given to the evolving nature of terrorist attacks, which many experts agree are changing. No one can deny that the Al Qaeda network has a clear intent to acquire and use weapons of mass destruction. Osama Bin Laden has said as much in multiple interviews and letters published through the media. Al Qaeda has even sought and received a religious ruling (fatwa) approving the use of a nuclear device against America. But at the same time, intelligence analysts warn of a global spread of the tactics developed in Iraq, Palestine, and Lebanon, tactics using smaller, but more frequent, improvised explosives attacks against "soft" targets. The natural targets of choice for such groups will be critical transportation nodes, power and other infrastructure assets, and various production and distribution facilities—the loss of which, in turn, could affect a great many businesses in unforeseen ways.

There also are numerous untold variables affecting the likelihood of any firm becoming a target that are quite simply impossible to quantify, such as the nature and identity of a firm's neighbors. In the greater Washington, D.C., area, for example, any number of seemingly benign suburban office parks and other sites house highly classified national security projects. An attack involving such a facility could easily cripple an unsuspecting neighboring tenant, even if only because of investigative and media attention following such an event. Such was the case when Timothy McVeigh's attack on Oklahoma City's Alfred P. Murrah building destroyed several nearby firms, including the daycare established in the same building.

Finally, the geographic footprint of a business matters a great deal in assessing the likely occurrence of discontinuous events, for if a firm has a particularly broad base or its value chain is particularly complex, events which might otherwise seem relatively unlikely may in fact occur with alarming regularity. General Motors, for example, discovered in analyzing their vulnerabilities that some portion of their value chain being affected by presumably "rare" events such as tornados or earthquakes in fact happens on a regular basis.[5]

> If a firm has a particularly broad base or its value chain is particularly lengthy, events which might otherwise seem relatively unlikely may in fact occur with alarming regularity.

Measuring Severity

The process of analyzing the severity of potential discontinuous events is similarly complex, and involves a determination of the assets that represent the most significant value to the firm as well as a measurement of the severity of various adverse events that could impact these assets (and, in turn, the overall enterprise). This task can be broken down into two constituent parts: identification of critical assets such as production facilities, buildings, vehicles and equipment, or certain distribution channels, and assessment of the adequacy of current protective measures. The most important assets can then be classified according to their criticality, sensitivity, and use within the firm. These facilities and products can then be provided additional protection through redundant power systems, development of alternate supply and distribution channels, and the securing of off-site emergency workspace for continuity of operations.

Vulnerability assessment teams should begin the overall asset assessment process by identifying the critical processes and key production nodes that are of the greatest value to the company. In particular, analyzing production and storage facilities, data acquisition and control systems, places for client interaction, and means and routes for shipping of raw materials and finished goods can provide valuable insight into the ultimate solution set.

Often, the first step is to examine existing documentation of key process flow documents such as the Material Requirements Planning (MRP) lists, computerized assets lists, dispatching and routing records, and other data maps depicting key components of the value chain. Easy solutions at this phase include identifying opportunities for the standardization of key parts, selecting multiple vendors to supply the specific items, and geographic (including international) dispersion of the means of production. It is during this process that firms typically discover the need to review, with further granularity, the processes by which their suppliers and distributors actually operate, for the value chain can only be made as strong as its weakest link.

The value chain can only be made as strong as its weakest link.

The next phase is to examine current security, process plans, and procedures in order to determine current and near-term vulnerabilities. The focus here is to protect essential equipment and infrastructure, for it remains incumbent upon the assessment team to ensure that no single accident, storm, or attack can endanger the enterprise as a whole, while also making recommendations in keeping with the business' overall profitability. Assessment teams will typically gather floor plans for key production facilities and any information on relevant security policies and procedures. They will look for any single points of failure, such as an over-reliance on single providers of key materials or on a single factory for production or assembly of needed parts.

Ultimately, solutions may make a case for dispersal of assets and certain production facilities, while other economic considerations support recommendations to maintain the status quo, consolidate assets, or use single-source suppliers to achieve economies of scale. The optimal solution will vary from case to case, but can include diverse alternatives. For example, a firm could choose to geographically disperse the assets in order to reduce the likelihood of simultaneous disruptions in locations performing the same operations or, alternatively, it could decide to co-locate critical assets and production and then add security and more survivable infrastructure to protect them. When it comes to security decision making, informed trade-offs can often provide the definitive answer.

Mapping Critical Infrastructure Dependencies

Looking at a firm's dependencies upon various critical infrastructures, including developing a risk map and drawing out the specific instances where key parts of the value chain intersect with the thirteen designated components of critical infrastructure is important. This analysis of the interconnected physical and logical relationships among various components (internal and external) to the organization itself forms the basis for understanding—and for avoiding potentially catastrophic events. For example, if a facility is served by a single bridge or road, or if it does not have at least two sources of electrical power, water generation, or telecommunications switches, then it faces potential risk from any significant disruption. Similarly, mapping out hidden relationships among the interconnected

When it come to security decision making, informed trade-offs can often provide the definitive answer. The right answer will vary by firm and industry, but only through deliberate examination of risks, threats, and mitigation alternatives will a firm be able to make the most rational choice for its operations.

infrastructures can identify how the closure of a seaport, for example, can delay the flow of parts, preventing the assembly and distribution of final products. Or, similarly this mapping can expose the ways a power disruption can render alarms inoperable because the phone system relies upon power to call the main switchboard.

While for the most part such infrastructure issues are larger than an individual firm can control—labor laws, the location and protection of bridges, and the overall security of telecommunications nodes are simply beyond the control of most companies—firms can examine the specific vulnerabilities related to those critical infrastructures and develop work-around solutions.

Again, the specific solution will vary by firm, but could include a myriad of activities such as constructing a secondary access road, establishing a contract to reserve excess trucking capacity, arranging for alternate power supplies and large back-up generators, or constructing support infrastructure, such as a water reservoir or temporary distribution facility. Expensive as these solutions may seem prior to the event, the time required to initiate these measures means they are often difficult or impossible to implement in the post-event phase.

Even when a firm doesn't have responsibility for the safety of the goods at a specific point of production, proper TSM and risk management strategies dictate that overall risk can be adjusted by addressing either likelihood factors or severity factors. Although many of the security aspects of a firm's value chain are beyond that firm's control, all consumers in a competitive market space have leverage relative to vendors. This means that they can contractually require certain provisions and security process improvements from their value chain or band together with industry partners to address significant shortfalls.

Assessing Findings

A risk matrix can also be a useful tool for conducting and presenting the risk assessment of a firm's security-relevant practices and processes. Relevant risks can be assessed and placed into three categories: critical, important, and acceptable. Each of these is then

Key Facility Utilities

- Emergency power and generators for fire sprinklers, water supply, and communications
- Fire pumps
- Telecommunications lines
- Storage areas for fuel, petroleum, oil, and lubricants
- Uninterruptible power supply (UPS) systems
- Climate control systems for products
- Stair, elevator, and utility shafts
- Distribution points for emergency power

color-coded as red, yellow, or green, respectively. Figure 4.3 depicts the typical approach used by the U.S. military, although the relative definitions for appropriate measures of severity of consequence and frequency of occurrence will need to be adapted to meet any firm's economic and other situation.

Implementing Mitigation Solutions

The final step in a logical risk management process is the evaluation and implementation of appropriate risk mitigation solutions. Significant factors include the predicted reduction in risk, the effect on daily operations, implementation and

Severity of Consequences			
Descriptive Category	Injury to Personnel	Equipment or Product Losses	Down Time
Catastrophic	Death	> $1 Million	> 4 Months
Critical	Severe Injury	$250,000 - $1 Million	2 weeks to 4 months
Marginal	Minor Injury	$1,000 - $250,000	1 day to 2 weeks
Negligible	None	$1,000 or less	< 1 day

* Adapted from the U.S. military's MIL-STD-882D

FIGURE 4.3   Assessing the severity of the consequences of an incident

Probability of Mishap		
Description	Definition	Time Frame
Frequent	Likely to occur repeatedly during defined time frame	Defined by each firm according to product or business cycle, such as 6 months, 1 year, 3 years, etc.
Probable	Likely to occur several times during defined time frame	
Occasional	Likely to occur during defined time frame	
Remote	Not likely but possible	
Improbable	So unlikely as to be assumed not to occur	
Impossible	Physically impossible to occur	

* Adapted from the U.S. military's MIL-STD-882D

FIGURE 4.4   Assessing the probability of an incident

maintenance costs, and unintended consequences like reduced efficiency or the shifting of risk to other similar or more valuable assets. Solutions may be as simple as increased employee training in regards to threat awareness and response plans, although it should also involve adding security and resiliency considerations to planned facility expansion, outsourcing, and similar long range planning activities, as discussed later on.

Risk Matrix						
	Likelihood					
Severity	Impossible	Improbable	Remote	Occasional	Probable	Frequent
Catastrophic						
Critical						
Marginal						
Negligible						

* Adapted from the U.S. military's MIL-STD-882D

FIGURE 4.5   Risk matrix

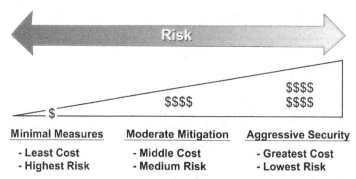

FIGURE 4.6   Spectrum of security alternatives

It is also important to recognize that the optimal risk port-folio is different for each company. A smaller firm may feel that in order to remain price-competitive it must simply accept the additional risk, while a medium-sized firm will take steps to mitigate the most likely operational threats but accept certain physical risks to reduce construction costs. A larger firm might decide to relocate to a less risky environment or better invest in its own physical security, while using its powerful market position to put pressure on its value chain partners to also better secure themselves. The right answer will vary by firm and industry, but only through deliberate examination of risks, threats, and mitigation alternatives will a firm be able to make the most rational choice for its operations.

Comprehensive, risk-based security requires new consid-erations for elements such as landscape architecture, land use, and site planning that make trade-offs among the firm's imperatives for aesthetics, risk reduction, and the support of normative business routines. It can also involve hardening select physical structures beyond regulatory minimums, in order to increase the likelihood of surviving a severe natural event, attack, or other disruption with minimal damage.

Conclusion

Many of today's best practices were developed by U.S. and other military services for worldwide use and are based on decades of security trial and error. These offer well-documented procedures for how to do things such as position guard posts, build effective fences, and design other structures to fulfill needed surveillance or security roles. The novel twist here is to begin to implement these best practices for addressing structural integrity and survivability, overt displays of security, and other measures while also focusing on solutions that create a return on the security investment. It may be helpful to remember that the weakest link in security is often the failure of individual people to properly carry out their duties. Metal detectors and package screening only work with well-trained attentive operators, and the strongest front door locks in the world are useless if the back-door is left open or other entrances are not monitored. For this reason security is a field where employee practices can play a defining role in the difference between a near miss and a debilitating blow.

Having explored the risk management process, the next chapters are devoted to specific risk mitigation and security enhancement measures that may be implemented to reduce or eliminate threats to certain physical and infrastructure assets.

Notes

1. Professor Dahl, "Quotes from Academic Papers by Graduate Students from Topics in Health Economics," *University of Minnesota Carlson School of Management on the Web*, 2002, <http://www.tordahl.com/academic/academic_1.html> (22 April 2006).
2. LCP Consulting and the Centre for Logistics and Supply Management, Cranfield School of Management, *Understanding Supply Chain Risk: A Self-Assessment Workbook*, 2003, <http://www.som.cranfield.ac.uk/som/research/centres/lscm/downloads/60599WOR.PDF> (8 April 2006).
3. Kevin Coleman, "Counter-terrorism for Corporations-Part II," *Directions Magazine*, 22 March 2006, <http://www.directionsmag.com/article.php?article_id=2126&trv=1> (1 April 2006).
4. Yossi Sheffi, *The Resilient Enterprise* (Massachusetts: The MIT Press, 2005), 33.
5. Yossi Sheffi, *The Resilient Enterprise* (Massachusetts: The MIT Press, 2005), 26.

"Your decision to be, have and do something out of ordinary entails facing difficulties that are out of the ordinary as well. Sometimes your greatest asset is simply your ability to stay with it longer than anyone else."

—Brian Tracy, *author and lecturer on managerial effectiveness, and business strategy*

Chapter Five

Securing Fixed Assets

One of the most significant imperatives for any business involves providing adequate protection for its means of producing and distributing goods. The harsh reality for many firms exists in the fact that if this capability is eliminated or degraded for any significant period of time, the entire business enterprise could be in jeopardy. TSM's four primary categories of concern, or the core areas that almost any enterprise needs to protect are: fixed assets, assets in transit, brand equity/goodwill, and human capital. This chapter explores the first of these concerns—ways to secure the fixed assets that enable the firm to function.

Fixed assets are typically defined as the more or less permanent assets that represent a significant investment and that are difficult to replace if rendered unusable. In the transportation industry fixed assets generally includes items such as property, buildings, factories, access roads and points of entry, and specialized production, maintenance, or transportation equipment. Securing such assets requires a "defense-in-depth" approach incorporating interlocking and overlapping security measures that, when combined, offer reasonable assurance that these assets will remain available for use during any disruptions.

Consider for a moment the consequences associated with having to replace fixed assets in transportation. Trucks can be purchased fairly easily on the open market, and tens of thousands of miles of railroad tracks are routinely replaced and repaired every year throughout the world. But what about replacing a ship? Building a container ship is no trivial matter. Whenever it has to replace a vessel, a steamship line often waits a year or longer to take delivery. Even shipyards that specialize in container shipbuilding, such as the Mitsubishi Heavy Industries yard in Kobe, Japan, generally need at least nine months from the time that the keel is first laid to finish construction on a 6,000 TEU vessel.

What if an airport was destroyed by a natural or manmade event? Imagine the scope and scale of a project to reconstruct a commercial airport. When the city of Austin, Texas voted to convert Bergstrom Air Force Base into a commercial use airport in 1993, the budget was $585 million. This was the biggest airport project in the U.S. since the construction of the Denver airport, and the project took four years to complete.[1] With this example, it's easy to see how starting out from adverse conditions surrounding a major discontinuous event could cost much, much more.

Finally, how could a state or country replace a seaport? In some cases, port facilities are the ultimate fixed asset—economic engines that can generate massive revenues for terminal operators and port authorities. Yet they are also assets that cannot be replaced or rebuilt easily.

For all these reasons it is critical that companies take a more active role in securing their fixed assets. Protecting critical fixed assets, access roads, physical structures, and vehicle and supply depots begins with the implementation of basic but overlapping security measures. The purpose of these efforts is two-fold: to deter potential attacks by becoming a relatively harder target than other nearby entities, and to be better prepared for withstanding or responding to any emergency that does arise by being able to rapidly reassert control over facilities and related assets. The TSM approach to securing fixed assets can be illustrated best through the lens of the four operational enablers: *best practices*, *situational awareness*, *training and exercises*, and *outreach*.

ENABLER #1: Best Practices Implementation

The primary focus of best practices for securing fixed assets revolves around the central tenets of access control and proper building construction. Access control starts even before the

Basic Physical Security Measures

· Secure all roof hatches and ensure emergency exit doors to prevent entry from the outside.
· Establish emergency plans and policies and procedures for evacuation and shelter-in-place contingencies.
· Properly light all building access points.
· Ensure adequate redundant power supplies for emergency lights and communications.
· Install an internal public address system.
· Use a credentialing system to control building access.
· Install security alarm systems and/or monitored CCTV.
· Design structure to resist progressive collapse and harden all exterior walls.
· Develop a secondary Emergency Operations Center.

physical entrance to the property. This is because most transportation facilities are dependent upon access roads and other points of entry that, if compromised, would result in an inability to ship and receive goods, effectively halting business operations.

For example, consider what would happen to operations at the Port of Miami if the Port of Miami Terminal Operating Company's (POMTOC) intermodal gate could not process inbound shipments. How critical is access control for Target Corporation's 1.5 million square foot East Coast import distribution center in Norfolk, Virginia? Or imagine the impact on operations if a significant portion of the over 9,000 tractors that J.B. Hunt operates were unable to access the depots that they utilize in over ninety cities in the United States.

Using a notional building as an example, effective access control should generally be designed in terms of concentric circles, with the outermost circle beginning at the perimeter of the property, the next circle encompassing the parking lot and main entrances, and the innermost circle continuing inwards until reaching specific floors or offices. In terms of basic design procedures, the primary considerations are ensuring good visibility and adequate lighting, both of which make high-risk areas such as back alleys, parking lots, and warehouses less susceptible to

FIGURE 5.1　Access control implementation

criminal acts while deterring unauthorized entry. Companies should also consider more extensive solutions, such as modifying angles of approach, regulating access to and control of loading docks, and using verifiable means of authenticating various customers and suppliers who visit the workspace or facility.

The next line of defense is the use of fences, guard gates, and, if appropriate, discreetly placed safety and landscape features such as curbs and holding ponds that control the flow of vehicular traffic. Other physical security aspects can include limiting or controlling the number of entrances and exits from the main roads, keeping perimeter gates locked, and optimizing for security the location, design, and lighting of parking lots. Some companies also employ guard dogs or electronic intrusion detection systems, but for other firms this is beyond the level of sophistication required.

Proper vehicle inspection is relatively low-cost, requiring minimally trained security guards to inspect all vehicles, including those of employees. Furthermore, just having a vehicle inspection station will deter some of those with dishonest intentions.

Vehicle inspection is another bedrock security function that should be standard procedure for most transportation firms, especially those who may face threats from terrorists motivated by religious, global, environmental, or other grievances. This is all the more important if a firm's value chain depends upon large production facilities or especially sensitive research and development laboratories that need to be secured. Proper vehicle inspection is relatively low-cost, requiring minimally trained security guards to inspect all vehicles, including those of employees. Furthermore, just having a vehicle inspection station onsite will deter some of those with nefarious intentions.

Few transportation facilities are as fortunate where natural barriers are concerned as the Florida East Coast Railway. The FEC rail ramp in Hialeah, Florida is almost entirely surrounded by the system of canals used in South Florida for water management. These canals serve as a kind of moat that encircles the railroad's critical areas and precludes vehicle and pedestrian access other than through designated gates. But even if an organization isn't blessed with existing natural barriers, a variety of affordable options are available, including the building of manmade lakes and the proper use of trees and sloping hills to prevent unauthorized vehicular access.

In addition, vehicle and pedestrian access routes should be designed with security in mind, and specifically with the goal of keeping potential car and truck bombs as far away as possible from the exterior of the building. This is paramount, because the dynamics behind the velocity of an explosion from a truck bomb dictate that the blast energy dramatically dissipates over even small distances, and even a relatively short area of separation can be the difference between a building that is scarcely damaged and one that becomes unusable. Barriers (also called "bollards") can be used to provide protective standoff distances between select points of entry and the facility itself, by blocking the main avenues of approach.

Some industries, including many hotels and airports, have now installed a combination of hardened versions of the more common items such as light poles, large tree planters, and

TSM Tools: Security Barriers

Security barriers are important components in controlling vehicular traffic and overall facility access. They may be manual or automatic and can include everything from movable gates and horizontal beams to large pilings, guardrails, or the concrete blocks known as Jersey barriers. An important distinction among such barriers is whether they are fixed, moveable, or expedient, as explained next.

Fixed barriers
Fixed barriers are heavy, permanently installed systems that provide maximum protection against unforeseen events. Examples include hydraulically operated retracting barrier walls, and can be either automatic or manual.

Movable barriers
Movable barriers can be moved from place to place. They may require heavy equipment or additional personnel for assembly after transfer. Highway medians, sand bags, sand barrels, or planters are just a few of the more typical movable barriers.

Expedient Barriers
Expedient barriers are temporary barriers created from materials or vehicles normally used for other purposes. They provide the least security because they are improvised from available materials and are less likely to be properly positioned. Buses, trucks, and cargo containers are some of the more common examples.

concrete water fountains or heavy benches, which now also serve a security function. If properly installed, such barriers can usually stop even sizeable cars and trucks, especially if they are traveling at manageable speeds. In addition, just as the previous solutions can control the flow of vehicles, a variety of innocuous features can also be used to manage the flow of pedestrians and cyclists.

Beyond the access control and external vehicular and pedestrian flow considerations, attention should be paid to assets in storage and/or awaiting transit which also must be well protected. The security associated with fixed assets for transportation mirrors the measures used across most industries, such as an array of video monitoring systems, entry detection alarm systems with magnetic contact switches, glass-break sensors to detect attempted entry through a window, and the use of safes and vaults to protect select assets.

Tropical Shipping is proud to exemplify these measures, declaring on their website that they keep all their warehouses well-lit, fenced-in and patrolled twenty-four hours a day by in-house security. They also constantly monitor these buildings with state-of-the-art alarm systems and closed circuit televisions, and they include locked and specially-monitored security cages for extra sensitive or high-value cargo.[2] Such alarm systems and security procedures are relatively affordable and constitute the baseline of what should be reasonably expected in today's security environment.

Additional security considerations might include installing additional, specific cameras and any necessary lights needed to monitor and record activity in the shipping area where trucks are loaded with the finished product. Cameras will, in large part, deter unauthorized use of the facility and its equipment. Additionally, cameras will deter employees with questionable backgrounds from working at the firm, as well as serve to document the truckload contents and load quality.

Finally, consideration should be given to the building structure itself, including the materials used and the design, at least in terms of structural stability. The most significant factor in designing survivable and resilient buildings is to select an outer material that is sturdy, avoiding the use of glass that can shatter and fragment from a variety of shock effects. If glass is unavoidable, consideration should be given to the use of special film that can be applied to glass to make it shatter-resistant.

TSM Best Practices for Access Control

At the Port of Vancouver, Canada, access control has long been a problem. Manning the entry control points cost nearly $125,000 (Canadian) per guard, but leaving them unmanned, in addition to the usual security and loss prevention concerns, meant long delays when locals used the port's roads to sidestep downtown traffic jams. Fortunately, the Port's Chief Security Officer, Graham Kee, has helped to design and implement a cost-saving, smart card access control measure that seems to yield benefits for all involved. The cards grant selective access to various parts of the 160,000-acre facility based on the level of permissions registered on the cards' control chip. What's more, when the port is operating at minimal security levels, a single guard can remotely monitor all twelve access points, and the automated gate system usually requires only that the trucker slow down, but not stop. This solution saves the trucking firm money because it saves time and the costs and penalties associated with additional inspections. It also increases efficiency by alerting loading crews at the terminals as to who is coming and when so they can be better prepared to load or unload immediately when the truck arrives. The system also saves the port money—to the tune of $125,000 (Canadian) per year for each of the 11 now controlled but unmanned gates. But, perhaps best of all, it saved an estimated $12 million (Canadian) by precluding the need to expand the roadway.[3]

ENABLER #2: Situational Awareness

The situational awareness requirements for fixed-asset security are based primarily on the need to know who is on a firm's property or in vehicles at all times, and the location of the firm's employees when they are making pick-ups or deliveries off-site. Understanding and tracking employee, visitor, and vendor movements while in or near critical areas can enable the firm to detect anomalies and to react before adversaries can do harm to the firm, whether it's theft of proprietary data or some more malicious act.

In addition, such knowledge will improve the firm's response functions because it will be able to determine employees' location, in turn more rapidly determining their status and any associated impact on their health or the firm's operations. Finally, it will enable the firm to more rapidly determine any loss of key personnel, which will enable it to focus additional resources for that asset or operation's resumption of business functions.

The vast majority of security incidents happen because criminals are able to take advantage of soft targets or weak security procedures. Understanding and tracking employee, visitor, and vendor movements while in or near critical areas can enable your firm to detect anomalies and to react before adversaries can do harm to the firm, be it the theft of proprietary data or some more malicious act.

Knowing the location of people—whether they're security guards, receptionists, or regular workers—is an imperative in today's business environment. Unlike monitoring the external perimeter, for the actual buildings that house critical human assets it's highly recommended that all firms consider the use of cameras, automated alarms, and card-key access systems on all employee entrances. These measures, in conjunction with keypad or other cipher locks, and the credentialing of employees and visitors with unique badges for positive identification provide a reasonable means for controlling both authorized and unauthorized access.

The Port of NY&NJ, for example, uses a combination of automated systems to securely manage the trucks and drivers that serve the port and terminal facilities each day and to synchronize their trips with dispatch. Using a uniform driver identification system called SEA LINK, the Port Authority is able to monitor truck movements and allow smooth entry and exit through marine terminal gates, cutting wait times and reducing pollution-producing engine-idling time. A single identification card allows authorized personnel access to any of Port of NY&NJ's marine terminals.

SEA LINK uses the port's Automated Cargo Expediting System (ACES) to transmit information about the drivers to terminal operators. This facilitates gate clearance and speeds up the drop off and pick up of cargo.[4]

Nonetheless, it must be remembered that the vast majority of security incidents happen because criminals are able to take advantage of soft targets or weak security procedures. The threat can come from anyone, from insiders such as a disgruntled former employee who retains their ID badge or a temporary worker who is stealing secrets to sell to competitors, to outside threats such as people posing as vendors or customers or illegally trespassing on private property. For this reason, overall security posture is extremely important and a firm's procedures should mitigate common threats by addressing a variety of considerations.

One of the most useful ways to track personnel movements is to institute a mandatory credentialing plan whereby all people—employees, vendors, and customers alike—have to receive and wear a custom photo identification card. This simple measure deters many would-be criminals because of the knowledge that the recording of their image makes them so much more likely to be caught, even if they escape. It also deters more serious crime because the location becomes a harder place to infiltrate relative to other locations, which decreases the likelihood of an incident.

ENABLER #3: Training and Exercises

The role and function of emergency response planning, training, and exercises for disaster-preparedness and business continuity is as imperative as it is obvious. People react much better to all manner of events when they are carrying out familiar tasks. The goal of training and exercises is both to identify weaknesses or oversights in existing plans as well as to improve people's reaction times if and when the firm needs to respond to a disruptive event.

Proper planning will help employees survive and recover more quickly from catastrophic events. Ensuring appropriate emergency planning accompanied by training and exercises is perhaps the single most effective measure any firm can take in order to increase the security of its assets. Regardless of the security solutions outlined above, the fact remains that accidents, attacks, and severe weather events do occur, and the only responsible decision is to be adequately prepared in advance. Training events and exercises are also needed because no plan is perfect and during a disaster with lives on the line is the worst time to find the mistakes in the plan or to be faced with decisions for the first time. Indeed, aggressive exercises can be a great investment because the response procedures will then become automatic, almost reflexive. Employee training and education efforts provide the baseline knowledge for all employees and managers, and tests, drills, and exercises are used to practice and improve upon established plans and provide practical experience under realistic conditions.

Although each type of firm will have specific needs to address, the primary elements of an emergency action plan can be defined as follows: an alarm or notification system, designated evacuation routes, known safe areas, an accounting of all employees, and employee training in maintaining critical firm operations. Many firms will also require that certain personnel be designated to stay behind (at least temporarily) to shut down specific equipment, look after certain high-value assets, monitor power and water supplies, and provide other essential services.

Employee awareness of how to respond to various events is a key determinant in reducing the toll in terms of lives and equipment.

A key determinant in reducing the toll in terms of lives and equipment is employee awareness of how to respond to various events. All employees should be trained in basic response protocols, and a designated subset of personnel should also receive specialized training in emergency first aid and other life-saving techniques such as cardio-pulmonary resuscitation (CPR). Common training issues include procedures for:

· Reporting incidents to authorities
· Evacuation and rendezvous
· Reacting to armed aggression, hostage-taking, or other life threatening situations

- Collecting information from verbal or telephoned bomb threats
- Handling of suspicious packages

Another important but often neglected aspect of security exercises is the concept of understanding and working within the limits of the amount of time required for an adequate police or security force response. After all, it does no good to detect an intrusion if the response force is too late or ill-equipped to deal with the problem at hand. Accordingly, good security requires the use of drills and exercises, and often, the hiring of a security company that can assure specific response times.

Militaries, as well as certain businesses, have been using simulations for centuries for practicing their plans in advance of actual events where they will be called upon to respond quickly and effectively. Corporate managers, shareholders, government officials, and other interested parties also can leverage training tools to provide a cost-effective way to evaluate return on investment related to security. Over time, exercises and simulations have proven extremely valuable in helping organizations to more ably respond to and deal with a whole host of disruptive events, including ones which do not resemble scenarios which the exercises included, but which nonetheless required decisive and informed leadership decisions to be made rapidly and under adverse conditions. The primary goals of the exercise are to reinforce the training that has been received and to identify any weaknesses in existing plans and protocols. Accordingly, training generally should precede the exercises in order to ensure that proper practices are first learned and then reinforced by the use of the simulated emergency.

As previously stated, there are three primary categories of exercises: limited, moderate, and full scale.

- Limited exercises (or "tabletop exercises") involve a group of participants walking through hypothetical events led by a control team that monitors their progress and injects additional problems as warranted in order to advance the scenario. Such exercises are somewhat artificial, and while they work through the overall logistical and other considerations of a crisis they typically minimize or avoid dealing with real-world practical complications such as vehicles that won't start, employees who can't be found, and the confusion that inevitably surrounds disruptive events. However, they are the least expensive of the forms of exercises as well as the least disruptive to normal business processes, and are a useful means for running through a variety of scenarios so as to vet overall plans and provide a high level education to the participants.

· Moderate exercises are still somewhat artificial but involve much more direct testing of portions of emergency response plans. Such efforts often include the actual deployment and testing of specific segments of the response plan to identify real-world complications, but often under a compressed and artificially-calibrated timeline. These exercises have greater impact than the limited ones described previously, but also tend to require more advanced planning and the participation of many more people.

· Full-scale exercises are, as the name implies, much broader in scope. They require a fuller response to the scenario's events and more fully test the actual real-world deployment of people and assets.

Selecting the appropriate depth of exercise requires an examination of the firm's objectives and the ability to make trade-offs between level of effort and the fidelity of the emergency response test. If the objective is to highlight the role of senior management versus actual on-the-ground realities then a limited tabletop may be appropriate, whereas if the goal is to test the specifics of the planned response to certain adverse stimuli it may require a more robust form of exercise. It should also be noted that for most firms, all exercises should be regularly scheduled and announced in advance in order to minimize disruptions to the flow of normal business (although holding unannounced exercises will provide a more accurate assessment of true preparedness).

One important side benefit of all the recent technological advances is that almost all facets of complex tasks, such as arranging realistic exercises, can now be simulated with a high degree of realism and variability. Regal Decision Systems, a company that specializes in computer modeling and simulation for transportation applications, offers a model to the industry that was designed as a simulation tool for evaluating operational, design, and policy changes related to maritime security initiatives. The use of such virtual exercises precludes the need for continually holding costly full scale exercises to achieve a high degree of verisimilitude, and also provides an excellent laboratory for carrying out the training that enables firms to examine different outcomes from various related scenarios and reactions.

In addition, recent developments in user-friendly, virtual, and three-dimensional (3-D) training environments have greatly improved capabilities in this realm. Building on classified work originally developed for the U.S. military and adding in the latest graphical interfaces and other simulation advances developed for the video game world, emergency response planners can now input building schematics, blueprints, or photographs to build virtual worlds that mirror existing structures. These ever more lifelike environments enable exercise planners to carry out increasingly more useful tabletop simulations and partial exercises at substantial savings in cost and time while maintaining the critical realism that ensures a valid learning experience.

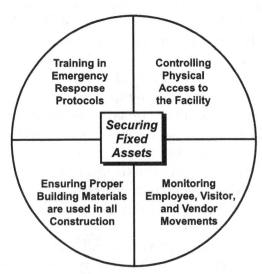

FIGURE 5.2   Securing Fixed Assets

ENABLER #4: Outreach

Appropriate outreach practices support fixed asset security in a variety of ways that, although difficult to measure in terms of specific dollar benefits, will become abundantly clear should an adverse event occur. These include the goodwill, cooperation, and personal relationships that have been created and which will be called into use during and after a discontinuous event.

First and foremost, good relations with neighbors and others in a firm's city or town can be critical for the early detection of abnormal patterns of surveillance or trespassers. It will also help to give a company the benefit of the doubt if something does occur and it believes it has or will have it under control soon. But perhaps most importantly, outreach includes getting to know the first responders—police, fire fighters, and others—that the company and its employees will rely upon in their hour of need. Indeed, first responders' familiarization with a company's facility and its layout could well cut through the confusion that makes the difference in whether or not the firm will suffer additional loss of facilities or even lives.

These relationships also bear more routine benefits, however, than just immediate emergency response. In October 2005, for example, Hurricane Wilma knocked out power to so many of the grade crossing signals and gate arms belonging to FEC Railway (who moves almost all of the intermodal freight up and down the Florida peninsula) that it caused a partial rail freight embargo throughout the state. Due in large part to their excellent working relationship with state and local officials, FEC was able to quickly restore rail service by enlisting the support of the Florida State Police. FEC Chief of Police Joe Walker worked closely with state troopers to co-ordinate manual guarding at the grade crossings that were inoperable, so that the trains could keep moving between Jacksonville and Miami. This action prevented a full embargo and helped towards speeding the entire region's recovery.

Outreach can also be important in less obvious ways. For transportation firms in particular there is often a difficulty in collective responsibility and shared control of certain common-usage facilities. This includes access roads and bridges and tunnels, but also even more basic features such as the ports that a terminal operator relies upon for their business, but which often in fact are controlled by individual port authorities or other similar quasi-governmental entities. In such cases, the best plan of action is to aggressively engage fellow stakeholders in the firm's concerns and emergency preparedness planning, in order to create collective efforts and produce maximum benefits from pooled resources.

Even in less than ideal situations, where a company's de facto partners in security elect not to engage with the same level of interest they are willing to put forward, open dialogue and outreach should help the company to gain some understanding of their partners' limitations and capabilities. Such knowledge will better enable the firm to prepare their own plans, and having the appropriate public persona and personal relationships also will help them have a voice in the aftermath of a disruptive event, increasing both their likelihood of rapid resumption of critical operations and of having some say in any temporary restrictive measures put in place to safeguard against additional incidents.

Applying the TSM Value Creation Model

The section below describes the specific tenets of applying the TSM Value Creation model to the task of securing fixed assets. In so doing it identifies the kinds of issues that firms should examine internally and externally, as well as the facets that should be examined by analysts, insurers, and regulators.

The appropriate public persona and personal relationships will give your firm have a voice in the aftermath of a disruptive event, increasing both your likelihood of rapid resumption of critical operations and of having some say in any temporary restrictive measures put in place to safeguard against additional incidents.

The Four Value Chain Vs: Visibility, Variability, Velocity, and Vulnerability

Visibility plays a key role in the defense of fixed assets—both in terms of the awareness of which assets are where within the firm's facilities (and which may be arriving soon), and in terms of who is doing what and why. Security also relies on ensuring minimal variability, in terms of the specific level of protection that is afforded, and although variations in patrol patterns do make it more difficult for anyone planning to do harm, this degree of tactical variability should not be mistaken for an excuse to allow operational variability in the overall security posture. Important variability metrics include adherence to access control standards, the construction of buildings from appropriately resilient materials, and minimizing the differential between protections afforded by the daytime and nighttime security posture. In terms of keeping goods moving through the value chain, velocity is also an important consideration, for TSM security practices can improve velocity through automated tracking and other solutions that simultaneously increase situational awareness and expedite handling to minimize costly storage time. These actions in turn lessen the period when those assets are vulnerable to theft or other events by virtue of their presence at a known, fixed location. Finally, vulnerability is perhaps the most obvious of concerns for securing fixed assets, and should be addressed through a combination of the access control, personnel screening, and other security measures discussed earlier in this chapter.

The Four Security Ds: Deter, Detect, Delay, and Dispatch

Deterrence is the most significant of the four security Ds, in terms of creating value while securing fixed assets, because the least expensive discontinuous event is the one that never takes place. However, detection of problems is a close second because if a problem does arise you need to know about it right away in order to begin to contain the threat, assess the damage, and potentially reroute incoming or outgoing goods. In other words, driving the criminal or terrorist away from a site by showing that it is a relatively secure target pays dividends through the events that don't occur, while rapid detection is essential to earlier and less costly mitigation. Delaying the onset of security breaches or the impact of natural events is also a significant form of security because it will, again, minimize the full extent of the event's damage and increase the time for preparing a response. Similarly, the timely dispatch of appropriately trained emergency, law enforcement, or legal and public relations response forces can also minimize the impact of the immediate event, as well as send the message that the firm is properly prepared for potential future events, which can bolster deterrence, build brand equity with the public, and increase analysts' confidence in terms of resiliency.

The Four Solution Set Cs: Coordination, Cooperation, Consultation, and Collaboration

Implementing the Four Solution Set Cs throughout the value chain plays an important role in a firm's ability to prevent, respond to, and recover from the adverse cascading effects of discontinuous events. It is through the combination of proper coordination and alignment of mutual business goals and imperatives that goodwill is created with all manner of external stakeholders. Similarly, continual daily cooperation builds the essential element of trust among your various value chain partners, while routine consultation provides the opportunity to resolve minor issues before they become large security vulnerabilities. Finally, collaboration on implementation of mutually beneficial solutions can go a long way towards building stronger relationships that will be critical in the face of the significant disruptions that accompany discontinuous events.

Conclusion

In summary, the knowledge of which risks still exist and the ability to act responsibly to plan feasible work-around solutions are core aspects of risk management. The goal of a comprehensive risk management program is not to avoid *all* risk, for that is an impossible task—and one that would bankrupt most firms even if it could be achieved. Certain risks simply must be accepted; indeed, this is a core principle within risk management, for the process is designed to elicit the proper perspective for analysis wherein risks must be mitigated, and therefore can only be managed. Since "forewarned is forearmed," however, it is often the process itself which is as important as the eventual risk mitigation measures put into place. This is true because the process of identifying weaknesses and evaluating alternative mitigation strategies highlights the areas of most significant concern.

Proper risk management also helps to avoid the all too common mistake of over investing in security or forgoing more reasonable alternatives by investing in more highly visible measures that do not add value to, or support the overall mission. One common example of mistakes in this area includes cutting the number of nighttime guards to pay for an increase in daytime security guards, leaving the nighttime guard force spread too thinly to effectively secure the workplace. Another is installing expensive, "real time notification" perimeter intrusion alarms for remote sites located in areas where no response force can reasonably respond in time to catch intruders. Such inefficient security expenditures are detrimental to the bottom line because they are inefficient, but they are also dangerous because they can give a false sense of security and lead to complacency regarding overall preparedness.

A related risk can occur even if a firm does invest properly in security: *if* the company fails to also implement proper value measurement controls. Without such controls, effective procedures can be instituted but a firm may then find that overall security and resiliency prove so good there is little theft and only minor disruptions. As a result, managers from other sectors or outside analysts and consultants may think that the firm should cut costs by reducing security and business continuity expenditures, when in fact those expenditures are the very reason the rest of the operations can be conducted so smoothly. Here again a core TSM principle comes to bear, that of pillar one: "Total Security practices must be based on creating value *that can be measured.*" If proper metrics are devised and utilized, then the overall benefits of the specific remedies can be properly atttributed and the beneficial practices will continue. In the end, that is what risk management is all about—creating reasonable trade-offs among risk mitigation alternatives, and finding effective ways to counter what risks remain.

Comprehensive risk mitigation to ensure the security of fixed assets involves teamwork—including collaboration with members of the value chain, industry partners (and competitors), and local or even national first responders and disaster recovery agencies. Creating a culture of security awareness, developing established threat recognition and response protocols, and increasing employee disaster response training are all important aspects of implementing proper fixed asset security. It also requires that thought be given to security in the design and layout of facilities and workspaces, to the securing of the perimeter, and to access control measures for all company property.

Business disruptions following "discontinuous events" can include everything from backlog and delay in value chain movements, to the loss of key production facilities for essential components, to the death or incapacitation of personnel or economic slow-down from national mourning. A company is responsible for finding a way to continue or resume operations as quickly and efficiently as possible. The best pre-event measure is the implementation of a rigorous emergency preparedness planning process accompanied by a variety of training and response activities. Whether man made or natural, when the next significant disruptive event occurs those companies that have implemented responsible risk management plans and properly addressed training and exercise concerns will find themselves much better off than their counterparts who chose not to.

> **Whether man made or natural, when the next significant disruptive event occurs those companies that have implemented responsible risk management plans and properly addressed training and exercise concerns will find themselves much better off than their counterparts who chose not to do so.**

Key Physical Security Actions

· Secure critical access and service roads, including bridges and tunnels
· Minimize the number of access points
· Ensure adequate lighting and video monitoring of all parking lots and sensitive areas
· Mandate biometric badges for all employees, customers and vendors, and create a security-conscious culture that challenges those without badges or walking in unauthorized areas
· Conduct regular training exercises with local police, fire fighters, and Emergency Medical Technicians (EMTs)
· Institute weekly intelligence/threat/updates, including review of best practices for security
· Incorporate provisions for families into emergency response planning

Case Study: Hutchison Port Holdings - Securing Fixed Assets

Value Chain Security Goals

· Leverage into CBP core principle of increased facilitation for legitimate business entities that are compliant traders
· Prevent illicit persons or substances, including weapons of mass destruction, from entering the United States in a shipment originating at HPH
· Create value by investing in security policies and procedures that will ensure uninterrupted access to U.S. markets.

Background: Hutchison Port Holdings

Hutchison Port Holdings (HPH) is one of the five companies that control over 80 percent of the world's terminal operating capacity. HPH handled approximately 42 million containers in 2005, and maintains operations in 35 ports throughout the world.[5] With such a large global presence, HPH realizes that the challenges and threats posed by discontinuous events can't be ignored by the maritime industry. Gary Gilbert, the Chief Security Officer, is on record as saying, "Each of the 42 million containers that went through our facilities around the globe was a Trojan horse." In an effort representative of the primacy of TSM's four primary enablers (implementation of best practices, knowledge management, training and exercises, and community outreach), HPH has invested aggressively in a security technology solution that has the potential to enhance end-to-end supply chain security by anticipating and preparing for contingencies associated with discontinuous events and potential container handling disruptions.

TSM Solution Set Case Study

HPH has committed to investing millions of dollars in security technology solutions, including the Integrated Container Inspection System (ICIS) offered by Science Applications International Corporation (SAIC). The ICIS system combines three discrete technologies for scanning loaded containers: gamma-ray non-intrusive inspection imagery, radiation screening portal, and optical character recognition, all in an attempt to identify illicit material, including components of weapons of mass destruction. This screening can be done without stopping a moving truck, and provides the terminal operator (and potentially its supply chain security partners), with valuable data that can help to efficiently and effectively identify high-risk containers. HPH is currently conducting an ICIS pilot project at two of the world's busiest marine terminals in Hong Kong. To date, the system has generated a database of 1.5 million images—all of which potentially contain information that have business and/or security value.[6]

Value Creation for Business Processes

There are several ways to create value within business processes:

- Aligns business and security processes with U.S. Customs and Border Patrol's publicly stated goal of providing incentives and benefits for international firms following prescribed security guidelines—including expedited processing of shipments
- Enhances end-to-end supply chain security while increasing supply chain performance, reducing the risk of loss, damage, and theft, and mitigating possible vulnerabilities due to terrorism
- Provides a way for security screening data to be shared with law enforcement authorities at the destination port
- Builds brand equity by establishing a presence as an industry leader in security practices
- Improves asset utilization by optimizing the cargo screening requirements and processes
- Enhances security for the firm's customers and workforce

Notes

1. Austin-Bergstrom International Airport, "Airport Project Fact Sheet," (2005), <www.ci.austin.tx.us/austinairport/projsumnr.htm> (20 April 2006).
2. Tropical Shipping, "Spotlight's On Security," <http://www.tropical.com/External/En/Press/TropicalNews/spotlight_security033006.htm> (25 April 2006).
3. Sarah Scalet, "The Mediator" CSO Online, June 2005, <http://www.csoonline.com/read/060105/mediator.html> (13 April 2006).
4. The Port Authority of New York and New Jersey, "About the Port Authority," 28 April 2006, <http://www.panynj.gov/> (29 April 2006).
5. HPH, "HPH – Introduction," (2001), <http://www.hph.com.hk/corporate/introduction.htm> (20 April 2006).
6. SAIC Security and Transportation Technology, *Integrated Container Inspection System (ICIS) To Enhance Container Security and Expedite Traffic*, October 2004, <http://www.saic.com/products/transportation/icis/ICIS_generic_white_paper_10-01a.pdf> (20 April 2006).

"From requiring new security language in contracts with their business partners to testing new technologies and ways to identify container tampering, it is private sector stakeholders that have been the innovators in securing their supply chains to protect their employees, customers and businesses."

—Jonathan Gold, *vice president of global supply chain policy, Retail Industry Leaders Association (RILA), testimony before the House Committee on Homeland Security, April 2006*

Securing Assets in Transit

Cargo container security ranks as one of the top concerns among transportation executives throughout the world. The shipment of containers may represent the greatest security risk in our cargo value chain. Over 70 percent of general cargo travels in containers today. In 2005, over 11 million loaded cargo containers, representing about $1.5 billion worth of goods daily, were imported into the U.S. alone. Growth rates of over ten percent are projected for the next several years, resulting in a projected volume of over 13 million containers per year by 2007. Unfortunately, this represents 13 million discrete opportunities for something to go wrong.[1]

The freight transportation business is about the movement of goods, and therefore the security of those goods while in transit is (rightfully) of paramount concern. The key to securing freight in transit is in implementing the right processes—gaining control from the first step by devising business practices that manage the security of goods or supplies throughout a firm's entire value chain.

> The key to securing freight in transit is in implementing the right processes—gaining control from the first step—if your value chain can assure that the cargo was secure when loaded and further ensure that it is not tampered with en route ipso facto the cargo will be secure upon arrival. Establishing control at the point of origin of each component significantly reduces the risk of a security breach and cuts down on the total cost of a move.

129

However, with the world reacting to the three change agents outlined in Chapter 1, effective Total Security Management of assets in transit must begin further down the line, sometimes even starting with a firm's third or fourth tier suppliers who may handle just one initial component of the product. This is critical because if the value chain can ensure that the cargo was secure when loaded, and further ensure that it is not tampered with en route *ipso facto*, it is a reasonable assumption that the cargo will be secure upon arrival.

Realigning security processes for freight has the potential to reduce reliance on random and incomplete inspections. At present, enforcement entities in the global transportation network are primarily concerned with spot checks that verify compliance with bills of lading and other regulatory schemes at pre-designated control points. This is similar to the way, before TQM, manufacturers would attempt to minimize variance by spot-checking final goods for quality assurance. Dr. Deming discovered that such an approach inherently accepts processes that allow unacceptable inefficiencies. It can also be expensive because it disrupts the flow of finished goods headed to customers.

As Professors Lee and Whang of Stanford University Business School write, "In manufacturing the way to eliminate inspections is to design and build in quality from the start. For supply security, the analogy is to design and apply processes that prevent tampering with a container before and during the transportation process."[2] Therefore, TQM-compliance would suggest that the solution to concerns with value chain security is to redesign the packaging and handling process as opposed to increasing inspections, which create inefficient costs in terms of bottlenecks, unpredictable delivery times, and demurrage losses.

Getting this process right is important because a disruption of the movement of containerized freight has immediate and wide-ranging economic implications for the global economy. When ports, rail yards, airports, and other key transportation nodes close, goods in transit can come to a standstill and global commerce is negatively impacted. When the 29 ports on the West Coast of the United States were closed to container traffic

for 9 days in September and October 2002 due to a labor conflict, the backlog of ships forced to anchor and wait to enter the Ports of Los Angeles and Long Beach grew to greater than 150. When manufacturing plants throughout the globe are unable to get the parts they need they are often forced to severely curtail or shut down production lines. Products such as food can be a total loss, if delayed, because of their time sensitivity for getting to the marketplace. Melons are imported into the United States in huge quantities at specific times of the year. As soon as the melons begin along their importation path, the clocks starts ticking. The time sensitivity of this commodity has a direct correlation to its value. Similarly, a 40-foot container of lobster tails, for example, can be worth as much as $250,000. But if the container is left idle and unattended for too long, and the temperature drops below a set level in the refrigerated container that carries them, the same load of lobsters tails can instantly become a total loss.

Whether it's for auto parts, cell phones, or melons, the global value chain is a complex system that is highly sensitive to disruptions. Automotive manufacturers typically realize revenues at the point of manufacturing, and delays in finished production can result in lost sales that deprive the firm of revenues needed to cover the industry's enormous fixed costs. If shipments to the auto plants are delayed, something as simple as a $100 spring can prevent the revenue generation for a $40,000 automobile.

Securing assets in transit has a number of bottom-line business impacts. If properly implemented, the risk of a major catastrophic failure of the global transportation network resulting from an act of terrorism can be reduced. Insurance costs can be driven down when pilferage and fraud are reduced,

Investment in state-of-the-art cargo security technology and monitoring solutions can provide significant return on investment—and often at bargain prices considering the value of the capital that could be lost by a disruption in the global container shipping. Tracking and monitoring cargo throughout the entire end-to-end movement increases visibility while establishing and preserving a proven chain of custody.

and corresponding claims decrease. The cost associated with compliance for a growing number of government regulations for freight shipment can also be reduced over time. Better control of shipment data, from end to end, also has the potential to improve compliance with regulations, reduce the number of inspections required at ports, and lower transit times. Inventory cycle times and stock inventory can also be lowered because the increased knowledge about the goods in transit allows for better planning throughout the entire process. Finally, improved customer service provides an opportunity to generate greater customer loyalty.

Processes implemented to enhance transportation make good business sense if they are designed to improve shipment visibility, which can in turn deter terrorism, impede smuggling, cut down on cargo theft, reduce the number of lost or misrouted shipments, enable authorities to identify fraud, and improve export controls. Investment in state-of-the-art cargo security technology and monitoring solutions can provide significant return on investment—and often at bargain prices, considering the value of the capital that could be lost by a disruption in the global container shipping.

Risk Mitigation for Goods in Transit

Protecting goods while in transit involves maintaining visibility and chain of custody through cargo tracking and monitoring, ensuring that the personnel handling the cargo can be trusted and have proper credentials, and safeguarding the physical security of the loading and unloading facilities by controlling access and providing adequate physical security. In addition, it is critical to be able to certify the security of the other partners in the value chain—the carriers who move its goods, the transfer facilities, the documentation professionals that handle the paperwork, the second and third tier suppliers, and the list goes on. The other essential element in a comprehensive approach to securing assets in transit is emergency preparedness, because even the tightest security cannot prevent all disruptions. It is crucial that a firm be prepared for all-hazards disruptions by planning for things such as substitute suppliers for goods, alternative launching or landing points for cargo shipments, recovery of information systems and established options for telecommunications, employee assistance, and family support, re-routing and alternate staging areas, and recovery plans for physical damage to the facilities.

Establishing control at the point of origin of each component significantly reduces the risk of a security breach and cuts down on the total cost of a move. The operational benefits of the process control improvements include fewer disruptions of shorter duration because effects of these disruptions can often be mitigated through early remedial action. Some disruptions can be planned for, and policies and procedures put in place to optimize the response. In general, more awareness regarding the movement of assets in transit equates to enhanced security and customer satisfaction.

The end-to-end TSM approach to security assets in transit can be illustrated through the lens of the four operational enablers: best practices, situational awareness, training and education, and outreach.

ENABLER #1: Best Practices Implementation

The focus of best practices in the movement of goods centers on the concept of end-to-end control of the movement of freight, necessitating the coordination of the security practices of suppliers, carriers, retailers, forwarders, and others. All players in the value chain web must be brought into the process to ensure that the entire movement is secure. End-to-end value chain visibility is a necessity, as it involves a comprehensive understanding of a firm's suppliers, shippers, carriers, and trade routes.

Tracking and monitoring cargo throughout the entire end-to-end movement increases visibility while establishing and preserving a proven chain of custody. To meet this objective a firm must first be able to guarantee the integrity of the loading and documentation process, including container sealing and shipment accountability procedures. Second, it must be able to monitor the shipment in transit in order to deter or detect tampering. Third, it should be able to provide complete, accurate, timely, and protected information for those who need the information throughout the movement. This process may include internal departments, external customers, other value chain partners, or government officials.

An estimated 20 million freight containers are currently circulating throughout the world, and roughly 7 million of them pass through U.S. ports every year.[3] Ensuring the security and integrity of those containers and their contents is the ultimate goal of any cargo security plan. Over the last several decades, goods movement has increasingly shifted from break bulk to the use of containers. Containers have delivered on the promise of efficiency, but they have also proven themselves vulnerable to theft and misuse by criminals. In fact, just after 9/11 in October of 2001, a 43-year-old Egyptian was found at an Italian port inside a modified container loaded aboard a German ship. The man's container had a small kitchen, a bed, food, water and batteries, and a cell phone. Similarly, on March 14, 2004, two Palestinian suicide bombers entered an Israeli port in a container and came out firing, killing ten port workers.[4]

Is the Company Prepared?

Have the firm's leaders undertaken a thorough risk assessment of the entire value chain and made adjustments based on the results? Some of these dangers could include:

- Weather or natural disasters.
- Infrastructure disruptions to power, water supply, natural gas, or communications systems.
- Labor strikes or other internal issues having to do with suppliers, carriers, ports, etc.
- Congestion and equipment shortages.
- For offshore outsourcing—political instability risks, currency exchange, rising level of economic well-being in many countries leading to higher labor prices, and competition for raw materials and supplies.

- Alternative supply chain networks, including other ports of entry or debarkation, other carriers.
- Economic factors such as rising fuel prices, inflation.
- Changing regulatory environment, locally and abroad.
- Employees improperly verified and credentialed.
- Terrorism.
- Availability of /competition for supplies during a time of crisis (for example, plywood, concrete, electric generators, medical supplies, water, trucks, and so on.)
- Adequate backup of data and the means to reestablish the system.

Does the firm constantly monitor its supply chain and incorporate best practices into daily business? These are some of the tools that might be used:

- End-to-end visibility of value chain.
- Using technology to track assets and monitor condition.
- Ensuring facilities are physically secure.
- Ensuring human capital is secure.
- Providing for availability of transportation assets/congestion.
- Tracking weather, natural disasters, and so on throughout the supply chain.
- Analyzing the financial condition of supply chain partners.
- Keeping abreast of political stability or economic shifts of markets in supply chain.
- Learning about new regulations or legislation.

Has the firm diversified its supply chain? The tools that can be used for this include:

- Using multiple suppliers and carriers in different markets to avoid the impact of a localized disruption.
- Providing alternative points of entry.
- Making alternative transportation routes and modes available.

- Providing flexibility for shifting to alternatives.
- Responding rapidly to spikes and slowdowns.

How well does the firm know its partners? Make sure that the firm:

- Requires disaster management plans from suppliers and carriers, and that they are discussed and updated regularly.
- Include partners in crisis training and exercises.
- Require a certain level of security from partners as a condition of doing business, and then monitor and audit regularly.

Does the firm have a crisis management plan in place? A good plan includes:

- Cross-disciplining crisis teams well and training them often.
- Establishing and testing a communications plan.
- Establishing processes and procedures for reestablishment or diversions in the supply chain.
- Making sure that all employees are aware of and trained in the plan.
- Cross-training employees to step in when needed.
- Using regular exercises to test and update the plan.
- Creating a mechanism to account for and assist employees, including alternative job sites.
- Backing up data facilities.

Especially since 9/11 the container has become the focus of national security efforts because of a perceived increase in risk. Many federal programs have focused on physical inspection technologies and protocols rather than changes in process controls. Perhaps this has to do with the fact that the private sector primarily "owns" the processes involved in moving containerized freight. It is important to note, however, that increased inspections would almost assuredly result in additional cost to the global transportation network, both in terms of finances and efficiency. The reality is that these inspections are also not likely to produce the intended results because they don't address the root flaws in the systematic process of securing goods. In contrast, the principles of TSM dictate that the focus should be placed on achieving process improvements that result in increased security. There are many proposed solutions for ensuring the security of the goods placed in containers and then preserving the integrity of a shipment while en route—many of those solutions are still evolving and may make their way into the realm of best practices over time. According to *Technology Review Magazine*, published by the Massachusetts Institute of Technology, currently, "...fewer than one percent of cargo containers—Pentagon cargo excepted—are (being) tracked with simple radio-frequency identification (RFID)."[5] However, industry is moving in the right direction. Hutchison Port Holdings, one of the world's largest terminal operators, and an RFID technology provider named SAVI Technology, announced in April 2005 that they would jointly invest $50M to build and activate an RFID network to track and manage container shipments.[6]

Tamper-proof seals have also been suggested as a potential solution for the end-to-end security challenge. In this case, all containerized cargo would be fitted with a seal to show

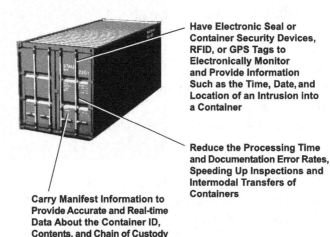

Have Electronic Seal or Container Security Devices, RFID, or GPS Tags to Electronically Monitor and Provide Information Such as the Time, Date, and Location of an Intrusion into a Container

Reduce the Processing Time and Documentation Error Rates, Speeding Up Inspections and Intermodal Transfers of Containers

Carry Manifest Information to Provide Accurate and Real-time Data About the Container ID, Contents, and Chain of Custody

FIGURE 6.1   Smart containers

that it has undergone and passed the first step in the security process when it was loaded. Affixing the seal signifies that the goods inside were certified to be secure and match the manifest for the contents.

The use of "smart containers" has also been proposed, and in several cases, even tested in real-world situations. A combination of mobile and fixed reader technologies can query sensor data throughout a container's end-to-end movement. RFID tagging and electronic seals have the additional benefits of allowing officials to monitor a container's movements around the world and provide data used to make informed decisions about which containers might require inspection before entering any given facility.[7] For example, containers can be fitted with electronic seals, Radio Frequency Identification Devices (RFID), and/or GPS devices. Such devices enable real time tracking and monitoring of the container, including data on intrusion detection, which, in theory, would provide notification if a container was tampered with while en route. General Electric has developed a solution that they call a Tamper Evident Secure Container (TESC). GE's partners on this project are Unisys and China International Marine Containers (CIMC), the world's largest manufacturer of containers. All of these solutions have

Smart Containers

Smart containers have specific characteristics, such as:

- Electronic seals or container security devices, RFID, or GPS tags for electronically monitoring and providing information such as the time, date, and location of an intrusion into a container.
- Manifest information for providing accurate and real time data about the container ID, contents, and chain of custody.
- A reduction of the processing time and documentation error rates, speeding up inspections, and intermodal transfers of containers.

something in common—they are security solutions designed to provide end-to-end security for freight in transit. Managing elements such as visibility and vulnerability for assets in transit are essential elements of the TSM approach.

Some multi-national terminal operators are taking the systematic screening of containers for business and security reasons very seriously. In Hong Kong, a demonstration project is underway to collect enhanced cargo screening for loaded containers entering the port to be loaded on ships. As a container travels through the port at about 10 miles an hour, the container passes through an optical character recognition portal that reads the container number from the side of the box, a radiation screening portal that screens for radioactive isotopes, and a gamma-ray technology that produces a non-intrusive inspection image of the inside of the container. In theory, by aggregating this screening data, anomalies and security risks may be detected more efficiently. This system has been designed to have minimal impact on the established flow of freight. Another added benefit is that the images of container contents can be shared and reviewed remotely on the receiving end. The cost is estimated to be about $7 per container.[8] This begs the question: "what is the cost per container for a shipper or carrier that is responsible for a shipment that ends up being the nexus of a catastrophic terrorist attack?". Is $7 a reasonable investment per container to try to avoid a disruptive event? One terminal operator believes so, and has invested significantly in this solution in an effort to enhance security and resiliency. Ultimately, this kind of return on investment analysis supports the Total Security Management approach to securing assets in transit.

Another consideration prompting better visibility is the increase in cargo reporting requirements. The U.S. Advance Manifest Rule (AMR) requires shippers to provide the bill of lading and cargo manifest for a container and its contents at least 24 hours in advance of shipment. The rule was put in place so that the documents could be reviewed well in advance of the scheduled arrival of the shipment, in order to allow for questionable cargo to be diverted or inspected as required.

TSM principles support the collection of such data as long as additional return on investment is realized by making the data more readily available to streamline inspection protocols and minimize delays in transit. To adequately ensure the security of the move, additional information could theoretically be shared electronically with every container shipped. This could include:

- Complete and concise cargo descriptions
- Container shipment origin—the name and address of the business where the container was packed
- Point of origin of the goods, including the country from which the goods are ultimately shipped and the name of the exporter or representative
- The name of the broker and the chain of custody of the container from its point of origin
- Name of the seller of the goods to the importer
- Name of the importer or representative and the ultimate purchaser of the goods

It is important to acknowledge that there are business issues associated with sharing this kind of data throughout the transportation community—related to privacy, proprietary issues, competitive concerns, etc.—that have not yet been reconciled.

Another capability that has the potential to provide both business and security benefit is dynamic re-routing. This is a fundamental business practice that has been perfected by trucking companies, steamship lines, and airlines alike. But the use of re-routing should not be limited to everyday operations. The ability to rapidly, efficiently, and effectively re-route freight in transit is an important continuity planning tool and can be the difference, in the case of disruptions due to a discontinuous event, between success and failure. Rapid and efficient adjustments to disruptions can result in significant cost savings; resiliency is good business.

Finally, establishing and verifying controlled physical access to cargo handling facilities can provide significant benefits to securing assets in transit. Technology implementations such as automated access control systems at truck gates, perimeter intrusion and detection systems, and sophisticated inventory tracking solutions all provide business process improvement and security benefits to users at ports, rail yards, airports, manufacturing facilities, warehouses, and loading docks. These solutions have the potential to be implemented by suppliers, shippers, and carriers throughout the value chain, providing a layered approach to security of the entire enterprise.

Several class one railroads in North America have chosen to use biometric fingerprint identification as a means to identify drivers entering and exiting intermodal ramps, and to validate their authorization to enter the facility. Canadian National and Union Pacific railroads use an ultrasonic biometric device, embedded in an ATM-style kiosk located in the truck lanes at selected ramps, which matches a driver against a pre-registered database in less than eight seconds. What's fascinating about these initiatives is that their roots were in business process improvement, not security. Both railroads needed a more efficient and effective way to identify and validate drivers. The solution, conceived prior to 9/11, ended up having significant and obvious security applications as well. Ultimately, by implementing this biometric technology solution, the railroads were able to save money, improve throughput, and enhance security. Several other railroads in the United States are considering implementing similar solutions.

In response to criticism that the industry was not implementing rigorous enough controls fast enough, the U.S. Department of Homeland Security announced in April 2006 that it would conduct background checks on dockworkers to ensure that they do not pose a terrorist threat. The names of the more than 400,000 port workers are being screened against immigration status databases and terror watch lists, but will not include, for instance, a criminal background check. In addition, some 750,000 logistics workers (dockworkers, plus truck and rail employees who have access to port facilities) will be issued

The Five Focus Areas for Securing Assets in Transit

· Establishing and maintaining a proven chain of custody—visibility, tracking, and monitoring of the cargo throughout the entire move.
· Confirming and accepting the security practices of the partners in the value chain—knowing the partners, and that the firm's level of security is only as good as theirs.
· Verifying the credentials of all personnel involved in the movement of the goods throughout the value chain.
· Establishing or verifying controlled physical access to the cargo and the handling facilities.
· Having emergency preparedness or crisis response plans to respond to disruptions that cannot be prevented, engaging and training personnel at all levels to know and follow them, and updating them often.

tamper-resistant identification cards.[9] TSM implementation supports initiatives that have the potential to improve business processes and enhance security. Any of these initiatives that can be implemented to lower risk, increase customer confidence, and minimize exposure to disruptions, work within the framework of TSM as long as concern is also paid to the need to create value through the new processes (such as through reduced theft, better insurance rates, increased throughput, or improved management of employees and their workday activities).

ENABLER #2: Situational Awareness

Situational awareness for securing assets in transit is focused on the information needed to track a firm's goods from the point-of-origin to their destination. Investment in technologies to increase end-to-end information sharing in real time can improve the ability to monitor for anomalies and to improve processes. By working toward a goal of optimizing network visibility into information about the location and condition of every shipment, knowledge of the likely choke points, and alternative sources of supply, a firm can better manage its assets and allocate its resources more efficiently. Additional concerns include knowing about other assets that can be brought into play, the length of time customers will be able to go without their goods, whether diversion of another shipment could solve a short term problem, and whether goods can be rerouted. All these factors can provide value in the event that a significant disruption occurs and are the keys to improving situational awareness.

End-to-end goods movement inevitably involves a large amount of data that must be protected throughout the movement. Electronic tags for example, can provide the means to capture and move the data with the shipment. GPS tags allow for global tracking of goods, while real-time information-sharing and electronic container seals allow real-time monitoring of container security. Use of these technologies also has the potential to provide both a security and business benefit. RFID technology is currently allowing many retailers to monitor stocks and automatically generate reorders. Wal-Mart, for instance, credits the RFID technology they required of their top suppliers for a 16% decrease in out-of-stock items—a move that had a direct impact on their bottom line and gave a boost to customer satisfaction. The suppliers benefit as well by knowing what and when to order and ship.[10]

Effectively managing and disseminating information on assets in transit has benefits beyond maintaining security. It can reduce inspection costs by reducing the number of containers actually requiring inspection, or expedite the flow through the documentation screening process performed by a customs organization. It may also contribute to reducing shrinkage by providing real-time information about the location and condition of a shipment: knowledge that may deter criminal activity, particularly from the inside. It can also reduce or eliminate lost or misrouted shipments and make delivery more predictable, in turn allowing for a faster restart of operations following an incident.

Knowing business partners and their security practices also heavily influences situational awareness. By gaining an understanding of the challenges that partners face, and then working collaboratively with them to address the challenges, a firm can contribute significantly to its overall security posture. Requiring minimum levels of security in contracts, and further requiring that higher standards be met over time, can form a foundation for continual improvement. As security best practices are refined and then shared with other members of the firm's value chain, the entire network becomes stronger and more resilient. Real-time measurement of security metrics shared with appropriate business partners can enable faster and more complete response to the disruptions that are bound to occur. Development of crisis contingency plans combined with regular training and exercises can also strengthen relationships and improve security.

There has recently been a push to develop global standards for information-sharing devices such as electronic seals, RFID tags, and GPS information systems. Some have suggested that this standardization could reduce costs and encourage the use of such devices. Just as complete end-to-end information is useful, and in fact necessary, to successfully implement Total Security Management, that same information would be

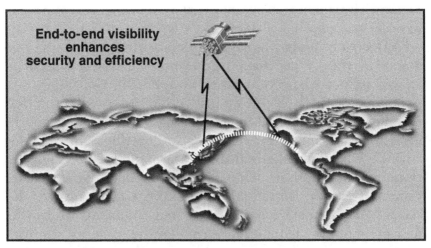

FIGURE 6.2   End-to-end visibility

equally as useful to those that intend to create disruptions. There is clearly a balance that must exist between information sharing to improve business processes and security, and the requirement to protect the integrity and custody of sensitive business information.

Establishing and maintaining secure control of assets while they are in transit also depends to a great extent on a firm's ability to validate the security practices of its value chain partners. The firm must know its partners because a firm's overall level of security is only as good as the level achieved by its partners ('weakest link in the chain' theory). Investigating and certifying value chain partners will enable the company to have confidence that the security plans in place for its portion of the value chain will not be eroded by poor security in one link in the chain. Beginning with an aggressive supplier selection process, firms can require complete information about business partners' security plans and procedures. This can include second, third, and fourth tier suppliers, intermediate carriers in other countries, and even any brokers or third party logistics suppliers who are engaged.

> Establishing and maintaining secure control of your assets while they are in transit also depends to a great extent on your ability to validate the security practices of your value chain partners. Requiring suppliers to demonstrate their commitment to security as part of a bid process can also provide an indication of a business partner's commitment in this area. You must know your partners because your level of security is only as good as theirs.

Site visits for verifying supplier's security claims and evaluating their security policies and procedures firsthand can be useful in this process. Requiring suppliers to demonstrate their commitment to security as part of a bid process can also provide an indication of a business partner's commitment in this area. Maintaining close contact, sharing information and best practices, and working collaboratively with business partners throughout the value chain can provide significant value to securing assets in transit.

ENABLER #3: Training and Exercises

The successful application of security measures is dependent upon educating the personnel involved and making certain that they understand the responsibility for ensuring the security of the cargo being moved. Training is essential at all levels throughout the value chain, both for the implementation of security measures and to ensure that all key players are prepared to respond if and when an incident occurs.

To ensure that assets in transit are secured, at a minimum, every person involved in the value chain must understand the role that they play in security. Involving stakeholders in training also provides an entry point into the process itself. Often workers actually doing the day-to-day work are aware of potential problems or weaknesses in security that are unseen by those developing the procedures. During training sessions, feedback from workers can be used to tighten security measures and improve their application. By employing these techniques, stakeholders can be co-opted into the process and may chose to invest personally in TSM implementation if they believe that their recommendations are valued and considered.

Training and exercises for emergency response plans can also represent significant value if the firm experiences a disruption. Employees who understand the part they play and practice their role over time are much more likely to make the decisions that will ultimately save the company money while preserving customer confidence. Business partners throughout the value chain can also be included in training and exercises. When a discontinuous event occurs, stakeholders are much more likely to be effective if they have worked together previously to solve a crisis. In some cases, protection of assets in transit can be entirely dependent upon the decisions that are made (or not made) by key personnel.

ENABLER #4: Outreach

Security of assets in transit requires the cooperation of both private and public sector entities. For TSM to create value, security must become a business driver and every partner in the value chain needs to be convinced that implementation of TSM initiatives represents value for them.

Appropriate TSM initiatives may vary significantly among firms. Some partners may even have difficulty seeing the value that it brings to them in terms of an immediate return on investment. One way that Tropical Shipping, a container line that serves the Caribbean, cultivates and enhances its relationships with business partners is to host tailored workshops on disaster preparedness and recovery. These workshops bring together various public and private sector stakeholders to coordinate planning and prepare for contingencies.[17] Additionally, Tropical Shipping hosts and maintains a web-based service that provides customers and business partners with current and actionable information on hurricanes. This site called: 'Hurricane Central," is available at: www.tropical.com, and provides recovery information and assistance to businesses (customers and value chain partners), governments, and individuals. During hurricane season Tropical employees, both in South Florida and throughout the Caribbean, also are trained and prepared to render assistance to their constituents. This transportation firm places a high value on contingency and business continuity planning, and is rewarded with strong customer loyalty.

A safe transportation system for all, free of preventable disruptions, may be little more than an abstract concept to the supplier of raw materials in a small South American town. The materials that they supply may eventually make their way into a finished piece of furniture for sale on the shelves of a major international retailer. Yet in reality, it is just as important for that small supplier to secure their piece of the business process as it is for the manufacturer of the final product. The campaign to make sure that all partners see and benefit from the value inherent in the application of Total Security Management must start at the top levels of management. A company should become an advocate both within its communities and throughout its value chain.

Companies can partner with local, state, and federal government organizations that impact the movement of freight to create value. These partners may include local and regional joint terrorism task forces, local law enforcement, departments of transportation, state departments of homeland security, and other government agencies including the Department of Homeland Security. The Hampton Roads community in Virginia, for example, seized the initiative almost immediately following 9/11 and formed the Joint Harbor Operations Center (JHOC). This facility was the first of its kind in the U.S. and provided a previously unavailable platform for the U.S. Coast Guard, U.S. Navy, and local port stakeholders such as the Virginia Port Authority to work together on security issues. The JHOC continues to serve as a command and control anchor for maritime security interoperability in the region, and has been called "one of the most diverse, sensitive, and valuable waterfront installations in the nation."[11] Additionally, The Virginia Port Authority Police Department, led by Ed Merkle, a former Coast Guard officer, has created significant value for the state and for VPA's customers by conducting community outreach and actively including internal and external stakeholders in the security planning process. Maintaining lines of communication, and involving appropriate public organizations in planning and training exercises can go a long way toward increasing security for assets in transit.

Sharing information with business partners and public sector stakeholders will improve the likelihood that they will support and endorse TSM initiatives. Once they recognize this value proposition the collaborative energy generated by working toward a common goal can result in valuable security and resiliency benefits for all concerned.

U.S. Government Value Chain Initiatives*

- 24 Hour Advance Manifest Rule: U.S. Customs Service implemented the Advance Manifest Rule requiring that shipping manifests must be submitted electronically to U.S. Customs for all shipments on vessels calling on U.S. ports 24 hours prior to loading from an overseas port. The manifests must include an accurate and complete description of cargo, full shipper and consignee information, and container seal numbers. This ruling requires closer coordination between shippers and their origin vendors and suppliers to ensure that they provide complete manifests.
- Automated Commercial Environment (ACE): U.S. trade processing system is currently being implemented to update the outmoded Automated Commercial System (ACS). It is intended to consolidate information gathering and automate border processing in order to significantly reduce time and cost and to enhance border security, which allows for more efficient collection, processing, and analysis of data. Further benefits include access by participants to their submitted data, improved communication, collaboration, and compliance within the trade community, as well as more efficient processing leading to expedited movement of goods across the borders. ACE should also be able to provide a platform for the sharing of trade information throughout the government agencies needing the data, eliminating redundant systems.
- Customs-Trade Partnership Against Terrorism (C-TPAT): A joint government-business initiative to build cooperative relationships that strengthen overall supply chain and border security. This public-private partnership between Customs and Border Patrol and the importing community is intended to motivate the private sector to implement

best practices in supply chain security through information sharing within the importing community. Businesses are required to ensure the integrity of their security practices and communicate their security guidelines to their business partners within the supply chain. The C-TPAT Security Link Portal is expected to enhance and improve the processing and communication for all C-TPAT participants and certified C-TPAT members.

- Container Security Initiative (CSI): A U.S. Customs program designed to ensure that all containers that pose a potential risk for terrorism are identified and inspected at foreign ports before they are placed on vessels destined for the United States. Stationed at foreign ports with the cooperation of the host foreign governments, multidisciplinary teams of U.S. officers from both CBP and Immigration and Customs Enforcement (ICE) target and prescreen containers to develop additional investigative leads related to the terrorist threats. The four core elements of CSI include using automated targeting tools to identify high-risk containers, prescreening containers at the port of departure before they embark for the U.S. (using advanced technology such as large-scale X-ray and gamma ray machines and radiation detection devices to speed up the screenings and avoid delays in the shipment process), and encouraging the use of smarter, more secure containers that ensure easy identification of containers that have been tampered with.
- Free and Secure Trade (FAST): A commercial process for expediting secure trade on the northern and southern borders of the U.S. by moving pre-approved eligible goods across the border quickly and verifying trade compliance away from the border. FAST-qualified trucks are eligible to move through FAST lanes equipped with RFID scanners, used to obtain information on the cargo content, driver, and tractor trailer without stoppage. Shipments

for pre-approved importers, carriers, registered drivers, and approved companies, are cleared into the country with greater speed and certainty, and at a reduced cost of compliance.

*Adapted from information available at the U.S. Customs website, www.customs.gov.

Applying the TSM Value Creation Model

The next section describes the specific tenets of applying the TSM Value Creation model to the task of securing assets in transit. In so doing it identifies the kinds of issues that the firm should examine internally and externally, as well as the facets that should be examined by analysts, insurers, and regulators.

The Four Value Chain Vs: Visibility, Variability, Velocity, and Vulnerability

All four of the Value Chain Vs are essential in establishing Total Security for goods in transit. Visibility is the cornerstone of situational awareness, and as such, effective tagging and tracking are the critical first links in knowing which assets are and are not where they are supposed to be. Variability must be controlled within this system, and although delays due to issues such as infrastructure problems, weather, inspections, and bill of lading errors will inevitably occur, it is essential to work with value chain partners in order to effectively control such costly variations in the transport of goods.

For the same reasons, velocity must be managed to ensure that the assets arrive neither too early nor too late at the next stop in their route, especially if the holding area for next segment involves unusually high risk. (Whenever possible, of course, process improvements dictate that higher risk locations should be avoided altogether, but sometimes routing is dictated by the point of production or is otherwise beyond the firm's control.)

Finally, although the vulnerability of goods can be difficult to manage once they are in transit, due to a variety of factors beyond the firm's control, one significant value creation solution is to ensure that they are not tampered with. If the goods were safe and secure when they were loaded, and you can verify that they were not tampered with in-transit, then they have a much greater probability of being secure upon arrival.

The Four Security Ds: Deter, Detect, Delay, Dispatch

The effectiveness of being able to deter stealing, tampering with, or otherwise affecting the safe and secure delivery of goods comes back to the web of layered security that a firm applies throughout its value chain. This includes matters such as the use of accountability procedures, security patrols, and verified background checks on employees and key value chain partners, all of which can make the firm a relatively more difficult target than firms that lack such protections. Similarly, the proper use of alarms, door seals, and content monitoring devices can help to detect intrusions and alert shippers and customers to the fact that a shipment has been compromised. Again, in many cases if the firm can take steps to delay the impact of the event, this also will save money by limiting the damage. Finally, dispatching appropriate response to incidents plays an important role in minimizing current impact as well as demonstrating the company's commitment to preparedness and resiliency.

The Four Solution Set Cs: Coordination, Cooperation, Consultation, Collaboration

Value chain coordination is important for all manner of transportation issues, but especially for the security of assets in transit, where by definition a company is reliant upon other firms and even specific individuals to protect the integrity of the shipment. Cooperation with these entities is paramount in terms of building trust and ensuring that the company will be kept abreast of any potential delays or other negative impacts.

Effective cooperation with value chain partners can also help to identify entities that should be high on the priority list in the event of a discontinuous event. In addition, seeking consultation with all manner of stakeholders can improve business processes by providing better understanding of the client's security and other priorities, and in the case of governmental relationships, by enabling the firm to anticipate future security requirements and regulations. Collaboration on mutually beneficial process improvements, such as standardizing asset control software or the sharing of sales velocity and related data to allow multiple parties to be more efficient and secure, has the potential to offer both a business and security return on investment.

Conclusion

Improved visibility and control are the cornerstones of securing assets in transit, and the best practices described in this chapter serve to support those goals. The intersection of TSM implementation and application of advanced technology solutions can be a powerful combination. Successful implementation of TSM requires the collaboration of all partners within a firm's value chain, from suppliers, to dockworkers, to carriers, to logistics providers. All stakeholders can potentially contribute to securing assets in transit and knowledge of all partners in the value chain, including their roles, security processes, and related business models, is critical to managing Total Security properly. A security breach that occurs early in the chain can negate all of the measures that may be taken further downstream. The opposite also holds true—effective security in the beginning and

> Knowledge of all partners in your value chain, including their security processes and business models, is critical to managing Total Security properly. A breach early in the chain can negate all of the measures that may be taken further downstream. The opposite also holds true—effective security in the beginning and middle stages of a shipment can be rendered meaningless by a breach at the end.

middle stages of a shipment can be rendered meaningless by a breach at the end. Metrics used to track the effectiveness of security processes throughout the entire value chain should be developed, applied, measured, and analyzed. The results of such effort would give all stakeholders a better ability to evaluate the company's relative risk preparedness.

Total Security Management, when applied to the challenge of securing assets in transit, can translate to measurable bottom line benefits for a firm. End-to-end tracking capabilities have the potential to streamline operations, reduce inventories, shorten lead times, improve predictability, and enhance security. By including all relevant stakeholders in the process of securing goods while in transit, a company's entire value chain can hedge its bet for success.

Case Study: The Target Corporation Securing Assets in Transit

Value Chain Security Goals

- Ensure that the supply chain does not have single points of failure and that multiple alternatives are available to continue operations.
- Enhance the end-to-end visibility and, by extension, the control of the supply chain for increasing reliability and security and for managing information.
- Ensure the resiliency of the entire value chain by understanding partners and by being certain that best practices and business continuity planning is in place and practiced.
- Use its ability to recover from an incident to also provide support and relief to the community at large during a crisis.

Background: The Target Corporation

The Minneapolis-based, upscale, discount, retail giant, Target Corporation, posted 2005 revenues of more that $52 billion.

Although the corporation dates back to 1881 with the founding of its parent store, Marshall Field & Co., Target's logistics and supply chain management are the embodiment of cutting-edge best practices. These include high asset visibility, advanced risk mitigation practices, and strong vendor relationships that enable the retail giant to deftly manage complex inventory and promotional mixes coming from more than 75 countries, through 23 distribution centers, and delivered to more than 1,400 stores in 47 U.S. states.[12] And they do it all through three import warehouses located on the U.S. West Coast, U.S. East Coast, and Ontario, Canada. Target didn't always have bi-coastal U.S. access and three U.S. points of entry, however. As recently as a few years ago Target had two import centers, both on the West Coast. According to *Logistics Management*, however, following the labor troubles all along on the U.S. West Coast in 2002, Target wisely opted to get around this potentially catastrophic single point of failure by using geographic dispersion and building the Suffolk and Ontario centers.[13] This discontinuous event prompted Target to reevaluate its supply chain and improve its resiliency.

TSM Solution Set

Target continually evaluates their security needs, and their current expansion plans include three more distribution centers and two additional import warehouses over the next two years.[14] Target also works hard at ensuring supply chain resiliency, including an analysis of the use of RFID tags to track their assets.[15] "Our goal is to make our supply chain incredibly reliable, and security is an important component of moving product from the manufacturer to our stores," says Diane Closs, VP of Distributions Operations.[16] But Target does more than just concentrate on their own facilities in order to ensure the viability and resiliency of their value chain. Target also actively engages its value chain partners in pre-event coordination and business continuity planning, working with internal and external partners to be able to respond to a supply interruption immediately rather than waiting and asking for their involvement in the aftermath of an event. As noted by Kelby Woodard, who at the time served as Target's Director of Supply Chain Assets Protection, "With a natural disaster, we can develop scenarios in advance that allow us to quickly get product moving to our stores. We employ other tactics to protect personnel, mitigate losses and support our retail stores."[17] Today Target is putting into action the kinds of programs that embody the four TSM operational enablers: implementation of best practices, situational awareness, training and exercises, and outreach to partners. The new programs will ensure end-to-end visibility of their supply chain, enhance resiliency by increasing the number of options available following a discontinuous event, and place Target in the forefront as a leader in the community during times of crisis.

Value Creation for Business Processes

- Eliminates choke points, enables recovery alternatives during disruptions through geographic dispersion.
- Improves security across the entire value chain through increased visibility of assets, the availability of more real-time information, and enhanced security.
- Promotes resilient practices with partners in order to strengthen relationships, heighten security awareness, and increase collaborative approaches to problems.
- Ensures strong, tested BCP plans for the survivability of the firm as well as giving it the ability to resume operations quickly. This allows firms to exert leadership in the community and provide relief and needed supplies in the aftermath of an incident, which also breeds goodwill and loyalty. Provisions in the BCP ensure that the firm can take care of their own personnel in times of crisis.

Notes

1. American Association of Port Authorities, Remarks of Christopher Koch, President & CEO of the World Shipping Council, 5 April 2005, <http://www.secureports.org/speeches/chris_koch_aapa_040505.html> (29 April 2006).
2. Hau L. Lee and Seungjin Whang, *Higher Supply Chain Security with Lower Cost: Lessons from Total Quality Management*, October 2003, <https://gsbapps.stanford.edu/researchpapers/detail1.asp?Document_ID=2273> (20 January 2006).
3. Sun Microsystems, "RFID Keeps U.S. Troops Well Stocked," *Sun Microsytems Services & Solutions*, November 2004, <http://www.sun.com/solutions/documents/articles/go_rfid_dod.xml, RFID> (29 April 2006).
4. Yossi Sheffi, *The Resilient Enterprise* (Massachusetts: The MIT Press, 2005), 145.
5. David Talbot, "Ports' Technology Failure," *Technology Review: An MIT Enterprise*, 26 February 2006, <http://www.technologyreview.com/read_article.aspx?id=16433&ch=biztech> (29 April 2006).
6. Savi Technology, "Savi Technology and Hutchison Port Holdings Establish New Company to Deploy RFID Network to Track and Manage Ocean Cargo Shipments," *Savi Technology Press Release*, 21 April 2005, <http://www.savi.com/news/2005/2005.04.21.shtml> (29 April 2006).
7. Ephraim Schwartz, "GE completes trial of smart shipping containers," *InfoWorld on the Web*, 11 January 2005, <http://www.infoworld.com/article/05/01/11/HNge_1.html> (29 April 2006).
8. Stephen E. Flynn, "Thinking Inside the Box," *Stanford Center for International Security and Cooperation*, 5 December 2005, <http://cisac.stanford.edu/news/thinking_inside_the_box_needed_to_secure_borders_write_cisac_professor_and_colleague_20051205/> (29 April 2006).
9. Lara Jakes Jordan, "Port Workers to Undergo Background Checks," *Yahoo News on the Web*, 25 April 2005, <http://news.yahoo.com/s/ap/20060425/ap_on_go_ot/port_security> (28 April 2006).
10. Debra D'Agostino, "Global Supply Chain Management," *CIO|Insight*, <http://www.cioinsight.com/print_article2/01217,a=173706,00.asp> (29 April 2006).
11. Harold Kennedy, "U.S. Navy Raises Barriers To Protect Base at Norfolk," *National Defense Magazine*, June 2002, <www.nationaldefensemagazine.org/issues/2002/Jun/US_Navy.htm> (29 April 2006). *See also* Greg Trauthwein, "JHOC: Eyes Wide Open," *MarineLink.com*, 8 June 2004, <www.marinelink.com/Story/ShowStory.aspx?StoryID=14627> (29 April 2006).

12. James Cooke, "Target zeroes in on import warehouses," *Logistics Management*, 1 March 2004, <http://sites.target.com/images/corporate/about/pdfs/target_history_timeline.pdf> (21 April 2006).
13. James Cooke, "Target zeroes in on import warehouses," *Logistics Management*, 1 March 2004, <http://sites.target.com/images/corporate/about/pdfs/target_history_timeline.pdf> (21 April 2006).
14. Target Corporation, *2005 Annual Report*, < http://investors.target.com/phoenix.zhtml?c=65828&p=irol-reportsAnnual> (29 April 2006).
15. Carol Sliwa, "Target Issues RFID Mandate to Suppliers," *Computerworld on the Web*, 1 March 2004, <http://www.computerworld.com/softwaretopics/erp/story/0,10801,90592,00.html> (29 April 2006).
16. Diane Closs, "Security Practices Safeguard the Supply Chain," *Michigan State University BROAD Business on the Web*, 2005, <http://bus.msu.edu/alumni/publications/broadbusiness/05/reality2.cfm> (23 March 2006).
17. "Security Practices Safeguard the Supply Chain," Broad Business School, at http://bus.msu.edu/alumni/publications/broadbusiness/05/reality2.cfm.

"No company wants to see its name on the front page of the Wall Street Journal being linked to a shipping container blowing up at a busy port or inadvertently selling sensitive goods to a known terrorist group. In addition to damaging the company's reputation, such news would also have significant financial repercussions... when a company announces a supply chain disruption, its stock price typically falls by almost 9 percent and $120 million or more of shareholder value is lost."

—Adrian Gonzalez,
Senior Analyst, ABC Advisory Group

Chapter Seven

Securing Brand
Equity and Goodwill

Companies, and even countries, expend a tremendous amount
of effort and resources to ensure that global customers have
a way to associate the services that they receive with a particular provider. As a result of this competition for market
share, brand equity is an important and valued asset for
every successful transportation firm in the world. According
to Deloitte Research, the concept of brand is a quantifi-
able asset that a majority of CEOs believe represents forty
percent or more of their firm's market value.[1] In the case of
transportation security, companies want their customers to
identify a safe and secure shipment with the transportation
entity that provided the service, hence building brand equity
with that customer. Governments similarly implement and
enforce security requirements to protect their citizens as well
as their economies—the theory being that customers will pay
a security premium to conduct business in a country that is
perceived to be relatively safer than others. In both cases, it is
the relationship between the provider and the customer that
creates associated brand equity. This chapter will explore how
protecting brand equity in transportation can facilitate additional business, as well as mitigate damages that can develop
as a result of a significant security incident.

Colors, Logos, and Calendars

How many times have you walked into the transportation office of a commercial carrier or into an import/export compliance office and seen a large, full-color calendar issued by a steamship line or a railroad with images of their ships and trains showcased over every calendar month? These calendars are an industry staple and most all industry professionals surely have been sent more than one such promotional item along the way. They decorate meeting rooms, cubicles, and hallways all over the transportation world. Have you ever wondered why transportation firms spend tens of thousands of dollars a year on these calendars?

Logos are permanently painted, bigger than life, on the side of almost every one of the 20 million or so containers crisscrossing the world. Ships and containers are painted in characteristic and easily recognizable colors that are associated with the company that owns them—who doesn't know that Maersk has arrived when the light blue container rolls through the gate? Railroads also have their signature colors and proudly display their logos on locomotives. (The Norfolk Southern black stallion rides the front of NS trains.) Trucking companies display their logos on the doors of their tractors and the sides of their trailers. Brand and name recognition is an important and integral part of the transportation business. The colors, logos, and calendars provide a way for transportation entities to develop a persona and for customers to identify with a particular transportation brand.

One definition of brand is, "a mixture of tangible and intangible attributes, symbolized in a trademark, which, if properly managed, creates influence and generates value."[2] This definition pays homage to the criticality of effective management where brand issues are concerned. A brand is not something that a transportation entity develops, possesses, and advertises without the expectation of a return on that investment. A brand must be properly managed in order to establish some level of associated value that is then assigned to the brand. Business schools around the world refer to this as "brand equity." And it is the value associated with brand equity in transportation that can be created—or destroyed—by decisions that are made regarding security.

Brand Equity in Transportation is Global

The business of transportation is a global enterprise, and as the trend toward increasing globalization of the business world continues, so follows brand equity. One does not need to speak Hindu, or Farsi, or even English for that matter to recognize the letters "CSX." Brand equity and goodwill can be created with or lost from customers throughout the world whose containerized shipments are moved by this or any other railroad.

The global nature of transportation has complicated the transportation firm's ability to manage customer relationships, however. Intermediaries such as brokers, freight forwarders, and third party logistics suppliers often buffer the direct relationship between shipper and carrier. Consequently, a transportation carrier may no longer be able to rely on personal relationships with customers to mitigate the damage that can be done to brand equity by a security incident. In order to develop and protect global brand recognition, firms must invest a significant amount of resources to create the results that ensure client loyalty and a positive perception of services. Businesses that invest in building brand and developing goodwill can be rewarded with repeat business and referral opportunities. In transportation, brand equity can be just as valuable, or even more valuable, than tangible assets such as trailers, ships, and locomotives. Without the positive brand association to bring in the customers, a transportation carrier is simply the owner of very expensive assets.

Imagine this scenario: a shipper in Hong Kong is notified that there has been a breach in the integrity of a containerized shipment bound for the Southern U.S., and that the incident occurred during the rail portion of the move from California to Arkansas. The first question the owner of the shipment will ask (a question that will color future transport decisions), is almost assuredly the name of the railroad company in control of the cargo at the time of the incident. And this is where the damage begins, because as soon as this shipper is able to associate the incident with the railroad's brand name the erosion of equity is realized. The railroad involved may not have a direct relationship with this shipper, and in fact may not even know that this shipper is associated with this particular freight movement—but

How Secure is the Brand?

Here are some questions to consider when looking at the company's branding:

Q: Is there a systematic way to inform others in the value chain when a security incident occurs?

Q: Does the company have a communications plan established that provides the organization with a way to standardize the message that is communicated to customers following a security incident?

Q: Does the company provide training for people in the organization that would help them respond to security inquiries from customers?

Q: Does the company conduct outreach and collaborative planning with its customers, in order to translate the value of its security initiatives and mitigate brand damage associated with a security event?

we can be sure that the shipper will begin to draw conclusions about the ability of that railroad to provide for the safe and secure movement of freight.

This hypothetical scenario is similar to what happened in March 2006 when the 5,500 TEU container ship Hyundai Fortune suffered an explosion and then caught fire at sea. The ship foundered in the Indian Ocean until the 27 crewmembers abandoned ship and were rescued by the Royal Netherlands Navy.[3] However, the media's primary focus in the press was on the steamship line, not the shipper of the load of fireworks that had been identified as a probable source of the explosion. The transportation firm, in other words, puts its name on the line with every shipment.

Transcending Interdependencies

Because global transportation infrastructures appear to be growing increasingly more interdependent, transportation providers have an opportunity to translate security initiatives, and the associated goodwill that is generated through the development of brand equity, throughout their value chains.

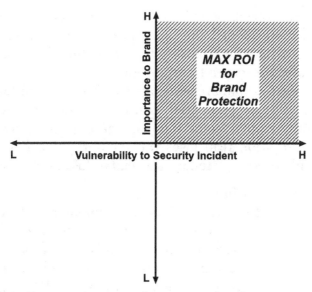

FIGURE 7.1   Return on investment for protecting brand equity

If a firm moves freight securely, it has the opportunity to translate that fact as a positive message throughout the value chain every time a shipment is successfully completed. The message is that the service is safe and secure. But is the firm taking full advantage of this opportunity to build brand equity and thereby create value?

Here's another way to look at this issue: at the start of a transportation shipment the parties involved in the associated global transportation network have (either implicitly or explicitly), made some level of commitment to ensuring that the shipment is safely and securely moved from origin to destination. This commitment requires an investment in resources. Brand equity can have its own capitalization rate and in order to effectively manage equity it should be monitored and managed in the same way that other aspects of a profit and loss statement are managed.

An Ounce of Prevention

A CEO or other person responsible for managing brand equity in a transportation network should ask this question: How much should the company spend to ensure that customers, and potential customers, do not associate our services with vulnerability and a high probability of loss? Is the answer $100,000, $1M, $10M, or more? To ask the same question another way, what is the cost associated with losing the reputation for being able to provide safe and secure transportation services?

Expenditures that are associated with security initiatives are often insignificant compared with the potential cost of loss of life or brand equity. As the following case studies demonstrate, if a transportation firm loses the reputation for being secure and reliable there is often an associated loss of goodwill.

Commercial Aviation

Following the terrorist attacks on September 11, the aviation industry experienced an immediate decrease in demand for passenger air services. In fact, this decrease in demand was so pronounced that several airlines faced the prospects of bankruptcy as their financial profiles and fleet operations were negatively impacted by the loss of consumer confidence (i.e., goodwill) in the security of air travel. Was this a complete anomaly or part of a series of events, illustrative of other goodwill issues in transportation, that have the potential to be repeated throughout the transportation industry as the direct result of discontinuous events?

Experts have estimated that the airline industry experienced financial losses, in the first week after September 11, of between $1 and $2 billion.[4] The effect on revenues was very pronounced. One industry analysis reported that total passenger aviation revenues showed a significant decline and 2002/2003 revenues were nearly 25 percent below revenues in 2000. Because the industry had experienced a discontinuous event that immediately eroded or destroyed the perception that air travel was safe and secure, the general public began to substitute other methods of transportation, which resulted in lost profits for all airlines involved. Most of these customers eventually returned to air travel, but not all.

The other element that drove many passengers away from air travel was the inconvenience related to ticketing and physical security policies enacted in the aftermath of the attack. One could argue that these initiatives were undertaken partially in an effort to restore goodwill, but they were also viewed as reactionary events by air travelers. The costs associated with these initiatives must also be accounted for in any discussion of goodwill, and it is important to balance the resultant positive impact that they may have had against the bottom line budgetary impact that they represent. The Air Transport Association has estimated that industry costs associated with new security processes and procedures, including insurance, may be as high as $3 billion.[5]

Railroad HAZMAT

A Norfolk Southern tank car derailment in 2005 resulted in a ruptured car and associated chemical spill (chlorine) in Graniteville, South Carolina. This particular accident produced a toxic cloud of poisonous gas that ultimately resulted in nine deaths and the evacuation of over 5,000 residents. That same year several major cities in the U.S. took action to attempt to ban movement of hazardous materials (HAZMAT) rail shipments in their jurisdictions. The connection between these adverse events and the citizens' demands that railroad traffic be rerouted is clear. And yet the fact is that 1.8 million carloads of hazardous materials (approximately five percent of total U.S. carloads) are transported by rail throughout the U.S. each year through most major cities. In 2004, 99.997 percent of these shipments were moved and delivered without incident—an impressive success rate by just about any measure. Furthermore, it is 16 times more likely for a hazardous release to occur from a truckload than from a railcar. Regardless, the railroads have received most of the attention on this issue.[6] It is the extremely rare event that threatens life, but it is that same event that could result in extraordinary additional costs to the industry.

To illustrate the potential damage these types of discontinuous events can inflict upon goodwill, we only need to examine the events that were initiated in 2005 related

> While statistically the rail industry can boast an exceptional safety record in HAZMAT transportation, even a few incidents have elicited strong reactions, especially as judged by the amount of media coverage. Goodwill is at stake because if not properly managed its loss can clearly result in a net negative business impact.

to hazardous rail shipments. A trend seems to have developed, primarily at the local governmental level. Throughout the U.S., local governments have attempted to prevent railroads from routing hazardous materials shipments through cities. Baltimore, the District of Columbia, Chicago, Cleveland, Philadelphia, and others have all made attempts to force the railroads to re-route trains carrying hazardous materials.

While statistically the rail industry can boast an exceptional safety record in HAZMAT transportation, a few incidents have elicited strong reactions, especially as judged by the amount of media coverage. Interest groups, such as environmental organizations, also may exploit an incident to demonstrate the danger of chemicals, and in so doing, elicit public and legislative support for its cause of limiting HAZMAT shipments near cities. This is bad for the business of the railroads and has resulted in a requirement for significant legal expenditures to defend the common practice of routing these shipments through urban areas. Goodwill is at stake in this case because if not properly managed its loss can clearly result in a net negative business impact.

Passenger Rail

Passenger railroads are not exempt from this kind of impact on their business. Following the 2003 and 2005 terrorist attacks in Spain and England, which resulted in more than 200 rail passenger deaths, significant changes were made in U.S. domestic railroad security in an effort to secure the system and prevent an erosion of passenger confidence. For instance, trash containers were hardened, law enforcement presence and patrols were increased, and employment of video surveillance and bomb-sniffing canines were rapidly increased. If the passenger (or freight) railroad industry in the U.S. is subjected to a major terrorist attack that targets their assets and customers the way that the September 11 attacks targeted the airlines, it is reasonable to anticipate a similar resultant loss of goodwill.

One can only speculate about the cost-benefit trade-offs that could have been made if some or all of the new aviation security initiatives had been implemented prior to, rather than following the September 11 attacks. Either way, there is clearly a direct correlation between goodwill and perceived level of effort that should not be ignored when examining security, demand for services, and profitability.

Affiliation with a particular transportation brand assumes some level of mutual trust and respect. When that trust and respect is eroded, the other aspects of the business relationship can suffer simultaneously. The cumulative effect of this loss of confidence can ultimately manifest itself in the decision to cancel services or transfer business to a competitor. In that case, to paraphrase an old adage, an ounce of prevention could be worth millions in saved business.

Hedge for Success

It is no coincidence that the busiest container port in the U.S., the Port of Los Angeles, displays an image of a port police boat on patrol in the banner at the top of their homepage (www .portoflosangeles.org).

This port is known to have placed a high priority on security initiatives over the last several years, and it seems clear that this image is being displayed to reinforce the message that the port recognizes security as a selling point and key differentiator within the maritime market.

Brand equity created relative to security preparedness and resiliency measures can be thought of as an insurance policy of sorts. Brand equity is not a "zero sum" game—a firm can gain or lose incremental amounts based on real or perceived status in the eye of the customer. Consequently, by building brand equity and developing as much goodwill as possible related to security prior to an incident, a firm may be able to mitigate the post-incident net loss of brand equity.

Business schools teach that brand equity, in dollar value terms, can be calculated by subtracting the value of a company's tangible assets from its market value. Conventional wisdom and standard accounting practices dictate that brand equity should normally be amortized over a forty-year period. However, the forty-year amortization schedule does not contemplate the accelerated loss of brand equity that can occur due to a security event.

Perhaps loss of brand equity can be indexed to the amount of effort expended on security prior to an event that results in a significant loss. A company would maintain brand equity following an incident if it: 1) maintained a security posture at or above a baseline level of security, and 2) demonstrated an acceptable level of security commitment to the industry at large. The perception (and ultimately the value determination by a customer) in this case could be that there was nothing more the company could have done to prevent the loss or that the loss could not have been prevented under any circumstances. In that case, the loss of brand equity should be dramatically lessened. Arguably, a company that cannot claim this posture or level of effort in the aftermath of a significant security incident stands to be blamed by the customer and is in a position to lose brand equity by nature of the association with an inadequate security posture (regardless of the actual relevance to the incident itself).

Consider two transportation firms, A and B. Firm A has employed an aggressive strategy related to security and has worked closely with its clients to communicate those efforts to its value chain partners, and the public at large, in order to develop brand equity in the marketplace. Firm B has only demonstrated a moderate commitment to security and as a result has a neutral reputation (or brand equity) related to security posture. Now assume that the same security incident occurs to both firms A and B. The net loss of brand equity for Firm A should be significantly less than that of Firm B, because Firm A had developed a strong stockpile of brand equity prior to the incident by making customers aware of security initiatives. Firm A's customers were also confident that Firm A had useable plans to recover quickly from the disruption. Conversely, Firm B had not accumulated brand equity related to security initiatives prior to the incident nor had it developed confidence among customers in the firm's ability to recover quickly. As a result, although the two firms experienced the same disruptive event, Firm A's value chain partners are more likely to stay with Firm A and give it a chance to recover, while Firm B has not received the benefit of the doubt from its customers and is at risk of a significantly greater loss of brand equity and market share.

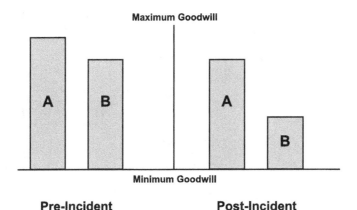

FIGURE 7.2   Benefit of Goodwill

Perceived Value and Security

Transportation customers essentially have two focus areas for determining value: quality and price. However, price is not where the effects of a major security event are absorbed. Another way to put this is that criminals and terrorists do not target price. What they do target and have a direct effect on is the quality component of the transportation value decision. We will focus our attention on the quality component of transportation services in order to explain how a security incident can effect a customer's perception of value.

The primary components of quality, at least where a transportation shipment is concerned, are on-time delivery and the integrity of a shipment from end to end (also known as the completeness of delivery). In other words, at the end of a shipment, two questions a shipper is likely to ask are: "Did it arrive on time?" and "Did it arrive intact?" If the answer to both of these questions is yes, it is very likely that the perceived value associated with that shipment is high. On the other hand, if the answer to one or both of these questions is "no" then it is very likely that the perceived value associated with that shipment is low.

Since security may be required to ensure both on-time and complete delivery, there is a direct integration of security and quality at this point in the value determination. If the perception

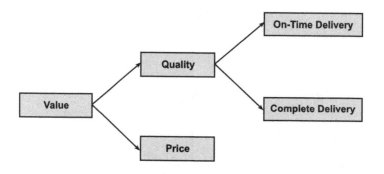

* Adapted from Bradley T. Gale, Managing Customer Value (New York: The Free Press, 1994), p. 29

FIGURE 7.3   Determinants of Perceived Value

of the value is poor, and the results that led to that determination (late or incomplete delivery) were also due to a security event, then the transportation firm is at risk of losing brand equity and having poor quality associated with its inability to provide a safe and secure shipment. If this value judgment is then extrapolated and applied to the brand itself, rather than being isolated to one particular shipment or incident, then the customer may begin to associate the lack of safety and security not to that shipment, but to the brand and the associated transportation provider in the overall.

Since security factors into both on-time and complete delivery, there is a direct integration of security and quality in the customer's value determination. The impact of a serious security incident can nullify years of investment in a single moment because the customer may associate a lack of safety and security not to a single shipment, but to the brand and the associated transportation provider.

Unfortunately, a serious security incident can do irreparable damage to a transportation firm's reputation, and has the potential to erode brand equity at an alarming rate. In the case of the transportation industry, where brand reputation can be entirely dependent upon a very limited number of service metrics, the impact of a serious security incident can nullify years of investment in a single moment.

ENABLER #1: Best Practices Implementation

A commitment to employing best practices in security requires a firm to make both a financial and a manpower commitment. The act of identifying, tracking, and employing these practices provides both an opportunity to create value by enhancing security posture, as well as an opportunity to create specific brand equity by pushing information related to the firm's efforts out into the marketplace. In order to create value, investors, financial analysts, employees, and customers should all be aware of the commitment that has been made, and the associated level of effort. Much in the same way that firms announce the hiring of new senior executive positions, mergers, acquisitions, and product development initiatives in order to capitalize on their position within the marketplace, companies should regularly make announcements concerning their implementation and use of security best practices.

Passengers moving by air are required to go through extensive carry-on baggage screening, and walk through metal detectors prior to boarding a flight. Yet passengers traveling by rail, even when carrying baggage, are usually required to do neither. A photo identification card is required to purchase a ticket at an AMTRAK counter, but tickets for the same train can be purchased online and picked up at automated kiosks without providing a photo ID before boarding. Some truck gates at freight terminals require fingerprint biometric verification of drivers, some require picture IDs, and others have no guards at all. It is not always easy to pinpoint exactly what constitutes the right level of security. But it is reasonable to expect that in terms of developing or forfeiting brand equity with a customer, there is a corresponding security posture that the average customer would expect the average transportation firm to maintain. This point represents the baseline, and is the starting point for the determination of best practices.

This baseline may be determined using a benchmarking process that analyzes the security posture and initiatives being implemented by similarly situated competitors in the market place. This process has been simplified by the consolidation that the transportation industry has experienced in the last ten years. Today, fewer than 20 steamship lines control more than 80 percent of the world's container shipping capacity; seven Class I railroads account for more than 90 percent of U.S. rail freight revenue and employ roughly 89 percent of all railroad employees; and five terminal operators control more than 80 percent of the world's terminal operating capacity. By maintaining a security posture and level of investment at or above this baseline point, a transportation firm may avoid being accused of being unprepared or having neglected to invest the requisite amount of time and effort in security if and when faced with a discontinuous event. However, if the firm chooses to neglect the benchmarking process by operating below a reasonable level, it runs the risk of incurring additional loss of brand equity when a customer conducts an analysis of its security posture.

ENABLER #2: Situational Awareness

In Chapter 2, situational awareness was introduced as a key asset that can be used to support the decision makers who have to set the priorities of the firm, and the same holds true regarding protection of brand equity. Using a systematic approach to collecting and disseminating information to various stakeholders and clients is much more effective than trying to develop and retain brand equity by employing ad-hoc or uncoordinated messaging techniques.

Managing the knowledge of what is and what could be is particularly important in an effort to create awareness for security initiatives and to derive some residual benefits from these investments. A firm that has an educated client base, that understands both the nature of the vulnerabilities and threats faced by a firm, as well as the specific actions that a firm has taken to address those challenges, will be much more likely to develop and retain brand equity. Over time, proactively managing security information should also support the overall security investment strategy. Security and resiliency considerations can play a key role in client relationships and are essential components that can drive a client's perceived value related to preparedness for discontinuous events.

Clients should be aware of what security efforts have been made related to securing fixed assets, assets in transit, and human capital. Awareness among all significant players in the value chain supports crisis management, but more importantly it also reinforces the message that the firm is taking action to avoid or mitigate transportation disruptions before they occur. This is a fundamental tenet of all TSM implementations (Pillar Two-*Total Security involves everyone throughout the value chain.*)

Situational awareness is especially critical to protecting goodwill if the company is forced to rapidly respond to a disruptive event. This is also the time when messaging and information transfer among stakeholders can be the hardest to manage, and is precisely why TSM requires a disciplined and deliberate approach to understanding and cultivating client relationships through situational awareness. Relationships with transportation intermediaries, carriers suppliers, and various other business partners can, in large part, drive the value associated with goodwill.

Finally, leadership on situational awareness at all levels in an organization can contribute significantly toward influencing a client's perception of the residual effects of a disruptive event. Responsibility exists throughout the value chain to ensure that accurate and timely information updates are made available to clients and stakeholders. Effectively managing key knowledge components, such as operational status reports and delivery estimates, can contribute significantly to an overall approach to securing brand equity.

ENABLER #3: Training and Exercises

Training can be a very effective way to close the gap between current and preferred security business processes. Many employees do not consider security to be part of their responsibilities, and may not feel compelled to maintain any level of awareness regarding security posture and initiatives. However, these

Training can be a very effective way to close the gap between current and preferred security business processes and it brings dividends in terms of demonstrable return on investment. The CEO of J.B. Hunt, which is one of the largest and most successful trucking companies in the U.S., made a corporate decision to manage training as a core business function and has publicly stated that the investment in training brings value directly to J.B. Hunt customers.

same employees will routinely come into contact with customers. Isn't it reasonable to expect customers to inquire about security issues with whomever they deal with on a day-to-day basis? What happens if there is a security incident and the first person that the customer comes into contact with following the incident has not been trained how to respond? Could this exacerbate the problem and cause additional erosion of brand equity? Could the customer infer from this encounter that the organization lacked a corporate commitment to security? At worst, training may help to mitigate additional loss of brand equity following a security incident. At best, an effective training program could not only help to avoid a security incident, but may create additional value associated with development of brand. As employees increase knowledge the customer's perception of the company's commitment total security can be enhanced.

In terms of demonstrable return on investment, it is not by chance that J.B. Hunt, which is one of the largest and most successful trucking companies in the U.S. and has over $2.7 billion in annual operating revenues, was named to Training Magazine's "Training Top 100" in 2005. The CEO has publicly stated that this investment in training brings value directly to J.B. Hunt customers, and Hunt has made a corporate decision to manage training as a core business function, employing 52 full time training professionals for its 16,000 nationwide employees.[7]

ENABLER #4: Outreach

Collaborative planning and outreach between transportation providers and their customers also can provide significant benefits in terms of generating or protecting brand equity. A collaborative planning process can serve to: a) convince a customer that the company is taking security seriously; b) provide tangible proof to a customer that the company is actively engaged in enhancing its security posture; c) provide the company, as a service provider, with valuable insight that may serve to mitigate loss in the event of a security incident; and d) provide both parties with a forum to discuss security concerns and engage in jointly developing solutions. This activity, combined with additional outreach initiatives to highlight the firm's security posture, can create significant opportunities for developing and protecting brand equity.

The Brand Equity Equation

A basic equation that illustrates the variables used by transportation customers to evaluate the service they receive and to correspondingly assign brand equity is as follows:[8]

$$\text{Service Level} = \text{Timely Delivery} \times \text{Complete Delivery} \times \text{Process Efficiency}$$

Assume that each variable can be assigned one of two values: one (1) equals perceived value, while zero (0) equals no perceived value. Note that this is a multiplicative and not an additive function because a value of zero for any of the three factors will result in an overall perceived Service Level of zero.

Also assume that any security incident that results in an incomplete or late delivery equates to "no perceived value," or a zero value, and is in keeping with many situations in today's transportation environment. Indeed, if a load of certain commodities is compromised the shipper often retains the right to simply abandon the entire load, and in some cases if retail shipments are received as little as 15 minutes late the consignee is entitled to reject the shipment.

Using this equation, one can see the effects that a security incident can have on perceived Service Levels. Even if the customer perceives positive value regarding timely delivery and process efficiency (for example, both assigned a value of one), if a shipment is compromised and/or destroyed by theft, sabotage, or some other cause and the complete delivery variable is assigned a zero, then in turn the entire transaction's services level rating equals zero. The same can be true of a shipment that arrives at its destination intact but is not delivered on time due to a security incident; the result becomes a zero because there is no perceived value in being 'just-too-late' in the 'just-in-time' world.

This brand equity equation is not intended to represent a scientific approach to assigning value, but rather as an illustration of how fragile the development and protection of brand equity associated with transportation services can be. This illustration underscores the importance of protecting brand equity. Fortunately, companies can take the following steps to protect their brand equity:

· Account for the value of brand and goodwill
· Assign responsibility for brand protection (roles and responsibilities)
· Involve employees and customers in the process
· Focus on the high risk/high impact vulnerabilities
· Use security successes to mitigate the impact of a negative event

Applying the TSM Value Creation Model

The next section describes the specific tenets of applying the TSM Value Creation model to the task of securing brand equity and goodwill. In so doing, it identifies the kinds of issues that a firm should examine internally and externally, as well as the issues that should be examined by analysts, insurers, and regulators.

The Four Value Chain Vs: Visibility, Variability, Velocity, and Vulnerability

A firm's reputation for delivering services and performing under difficult circumstances is intimately tied to its perceived brand equity. Supply chain visibility is a powerful tool for businesses to keep customers fully informed of the status and condition of their shipment. Having end-to-end visibility and being able to provide that to a customer almost real-time has become a hallmark of some of the most successful shipping companies. FedEx and other overnight shipping companies have based entire marketing campaigns on being able to tell customers exactly where their package is and when they can expect it. The transportation firm has opportunities throughout the shipment process to re-assure a client that their shipment is secure by being able to provide shipment visibility. If the firm can offer this service it should similarly gain a competitive advantage over others in the marketplace that cannot.

By using technology the firm also can better control variability in the quality of services provided. Consistent monitoring and mitigation, as well as keeping the customer informed of any known variability, will ensure that the company builds a reputation for doing everything in its power to minimize variability. Often

Timing is everything in today's economy and failure to deliver on time can do irreparable harm to your brand equity.

the goodwill that is created when a customer has confidence that a firm has initiated appropriate processes to prevent variability is what helps keep that customer from leaving if adverse events do affect product flow. Velocity is another important and highly visible aspect of value chain operations and is often the one that presents the most problems for firms. Timing is everything in today's economy and failure to deliver on time can do irreparable harm to the brand equity. Reaching out to partners and integrating asset-tracking tools to ensure full visibility will help the company better manage velocity, as will analysis of avoidable choke points where there are routinely-experienced and undue delays. Finally, since no firm is impervious to vulnerability, a company needs to understand and manage its exposure and keep its customers informed as to the relevant risk management measures it uses to protect their shipments and assets. In the end, a company with a well-documented and useable risk management strategy will be in a better position to recover quickly, to retain its customer base, and potentially even to gain customers who leave the other firms that are less able to prevent or manage disruptions.

The Four Security Ds: Deter, Detect, Delay, and Dispatch

The Four Security Ds present opportunities for a firm to communicate to the value chain partners and customers that it has taken the proper steps to ensure an appropriate security posture—and this fact can be used to enhance brand equity. Employing visible security measures, such as cameras, access control gates, and barriers to deter and detect security disruptions and publicizing initiatives such as the use of smart containers and biometric identification systems not only assure partners that the firm is engaged in proactive security, but it also deters would-be thieves and terrorists. The ability to discover potential disruptions in time to invoke a contingency strategy that ensures success can be used as a differentiator to attract and maintain business. Delay also is important because the ability to temporarily hold off a threat while devising an appropriate solution is absolutely critical in terms of protecting the firm's reputation for safe and secure service. Disruptions and threats of disruptions must be dealt with in a decisive and efficient manner to avoid negative public perception. A company with a reputation for decisive dispatch of disruption response also will appear better defended against such threats and be an unappealing target in the future, further enhancing its image as a security conscious firm.

The Four Solution Set Cs: Coordination, Cooperation, Consultation, and Collaboration

The key to building and maintaining brand equity with respect to security depends significantly on coordination and cooperation throughout the value chain. The ability to coordinate with partners and customers ensures the secure operation of business processes. Firms that invest the time and effort to coordinate security, risk mitigation, and business continuity plans with those who might be impacted or those who can have an impact will have a higher probability of success, and success leads to satisfied customers. Cooperation creates value in the form of goodwill with customers, partners, and community stakeholders.

Consultation similarly serves to enhance the reputation of the company by positioning it as a trusted player in the global community. Attending industry conferences and exchanging best practices with counterparts not only strengthens the security posture of the whole industry but also showcases the efforts of the firm in a public forum, which again pays dividends in the form of increased brand equity.

Conclusion

By recognizing business imperatives related to security in transportation and establishing aggressive, corresponding goals regarding security posture, a transportation firm can create and protect market place brand equity, and consequently create value. By measuring and managing brand equity related to security, senior management can set a transportation firm apart, distinguish it from competitors in the market place, and provide customers with additional reasons to continue or enhance their existing business relationships with the firm.

Essential elements that build brand equity in transportation and can be affected by a security incident include: shipment speed and accuracy; condition of freight at delivery; shrinkage of freight in transit; and efficiency of service. In order to ensure continual progress in developing brand equity among customers, employees at all levels should be trained and empowered to be part of the security solution.

Brand equity in transportation can represent a significant portion of a firm's asset portfolio, perhaps even more valuable, in some cases, than the tangible assets. In an effort to create value, officers of a corporation have an obligation to protect brand equity much as they are obligated to protect other assets of value such as financial capital and employees. TSM can support an effective corporate risk management strategy designed to protect and develop brand equity by minimizing exposure to significant security risks, involving everyone throughout the value chain, and aggressively managing post-event recovery initiatives.

Case Study: The Home Depot—Securing Brand Equity and Goodwill

Value Chain Security Goals

- Driving supply chain improvement, establishing return on investment, and increasing service levels through implementation of command and control technology solutions.
- Optimizing day-to-day efficiency, productivity, and resiliency through employment of state-of-the-art automated load planning, inventory optimization, communications, weather management, and asset tracking systems.
- Efficiently scheduling store delivery appointments, maximizing storage yard capacity, and minimizing turn times.
- Leveraging data, identifying trends and improving performance throughout the supply chain network.
- Employing command and control resources that can ensure uninterrupted service to U.S. retail markets.

Background: The Home Depot

The Home Depot is the nation's leading diversified wholesale distributor, and currently has 2,040 retail home improvement stores throughout North America, Puerto Rico, and the Virgin Islands. Their client base includes residential, commercial, infrastructure and industrial construction, and maintenance consumers. The Home Depot realized that the challenges and threats posed by discontinuous events could not be ignored by the retail industry. In an effort that combines all four of the TSM operational enablers (implementation of best practices, knowledge management, training and exercises, and community outreach), The Home Depot has invested aggressively in business continuity planning assets. By anticipating and preparing for contingencies associated with discontinuous events and retail distribution disruptions, the firm continues to develop and protect goodwill among its public and private sector customers. The Home Depot ranked

12th among major companies in Harris Interactive's 2005 Reputation Quotient, an annual, 600-person survey of "reputation-related categories," including Quality Service and Social Responsibility.[9]

TSM Solution Set

The Home Depot has made a significant investment in its Store Support Center. This is a centralized command, control, and communications facility that has been specifically designed to support a flexible and responsive supply chain, while minimizing business disruption and optimizing The Home Depot's ability to respond to and recover from discontinuous events. The Transportation Management Solutions Team has implemented a Transportation Planning and Execution (TP&E) initiative that facilitates continuous incremental improvement in supply chain performance. This includes enhanced inventory planning, optimization of asset utilization, and streamlining distribution channels.

Each Home Depot store is equipped with a computerized point-of-sale system, electronic bar code scanning system, and a UNIX server. By linking these systems back to the Store Support Center, and leveraging the data with the weather, asset management, and inventory control systems, the firm is able to employ an extremely efficient and responsive store-based inventory management, rapid order replenishment, and optimized transportation delivery environment.

Value Creation for Business Processes

· Enhances end-to-end supply chain security while increasing supply chain performance. This reduces the risk of loss, damage, and theft, and mitigates vulnerabilities that may arise due to terrorism.
· Improves communication between facilities managers and vendors during emergencies and allows managers to be more informed and involved.

- Enables rapid and effective response to changes in retail demand related to discontinuous events such as unusual or delayed weather patterns.
- Builds brand equity by establishing a reputation as a reliable and resilient wholesale distributor of retail building supplies.

Notes

1. Deloitte Research, *Prospering in a Secure Economy*, 2004, <http://www.deloitte.com/dtt/cda/doc/content/DTT_DR_ProsSecFull_Sept2004.pdf#search='Deloitte%20Research%20prospering%20in%20the%20secure%20economy> (1 April 2006).
2. Tutor2u Limited, "Interbrand Branding Consultancy," <http://tutor2u.net/> (22 April 2006).
3. Eagle1, "CTF 150 rescues 27 from burning ship off Yemen," *23 March 2006*, <http://eaglespeak.blogspot.com/2006/03/ctf-150-rescues-27-from-burning-ship.html> (25 April 2006).
4. J.N.Goodrich, "September 11, 2001 attack on America: a record of the immediate impacts and reactions in the USA travel and tourism industry," *Tourism Management*, 23, no. 6 (2002), <http://www.sciencedirect.com/science?_ob=ArticleURL&_udi=B6V9R-46081GP-4&_coverDate=12%2F31%2F2002&_alid=393521211&_rdoc=1&_fmt=&_orig=search&_qd=1&_cdi=5905&_sort=d&view=c&_acct=C000050221&_version=1&_urlVersion=0&_userid=10&md5=d0d782775f9a636886ee27a74c15e461> (13 April 2006).
5. Robert Poole, *Air Travel Two Years After 9/11*, 11 September 2003, <http://www.rppi.org/airtravel911.shtml> (April 15 2006).
6. American Association of Railroads, "About AAR," 2006 <http://www.aar.org/> (25 April 2006).
7. Securities and Exchange Commission, "Form 10-K for J.B. Hunt Transport Services, Inc.," 3 March 2005, <http://www.sec.gov/Archives/edgar/data/728535/000110465906015770/a06-1990_110k.htm> (24 April 2006).
8. Martin Christopher, "Homepage," 28 April 2006 <http://www.martin-christopher.info/about.htm> (28 April 2006).
9. The Home Depot, "About the Home Depot," <http://corporate.homedepot.com/wps/portal/> (15 April 2006).

"A man's labor is not only his capital but his life...To utilize it, to prevent its wasteful squandering...this surely is one of the most urgent tasks before civilization."

—William Booth,
founder of the Salvation Army

Chapter Eight

Securing Human Capital

Transportation firms often come face to face with the significant business implications of the discontinuous threats that natural and man-made hazards can pose to the survivability of an organization's workforce. Jurisdictional issues, poor training, inadequate communications, and lack of collaboration among those responsible for mitigating the impact of discontinuous events can further complicate these issues. Recent events involving natural disasters and terrorist strikes, both inside and outside the U.S., have illustrated and underscored the vulnerabilities that exist related to securing human capital. Globalization has created a multinational work force, in some cases lowering the cost of skilled and unskilled labor, and increasing organizational vulnerabilities because the value chain now stretches vast distances across both physical and cultural gaps.

And yet transportation is an industry that relies almost entirely upon people, making its human capital transportation's most valuable asset. In part this is because the industry is so fragmented that goods change hands many times between their origin, the factory, the wholesaler, the retailer, the carrier, and the end-user. It is the people in the transportation industry that make bookings, load containers, sort packages, complete export documentation, sail ships across the oceans, load railcars, and route trucks. Ships don't navigate themselves; trains don't get to their ramps on time without conductors; and planes do not get

187

loaded with cargo by machines. People are involved in virtually every aspect of transportation. It follows that human capital must be a critical component in creating value, and it is reasonable to argue that an effective strategy for managing this capital must include a comprehensive security component.

The Company's Greatest Asset

Most business owners agree that it is extremely difficult to recruit and retain talented employees. It is usually an expensive and time-consuming process for employers to find employees with the right "fit"—the concept that the personality, education, and experience that the company requires are aligned with the corresponding professional objectives of the employee. The resources that a company expends in pursuit of employee talent must be immediately reinvested whenever an employee leaves the company. High turnover rates can be especially costly to employers because it can takes months for a firm to find the right people for a job, and even longer for a new-hire to settle in. In fact, it takes the average employee six months to adjust to a new job. According to Hamilton Beazley, chairman of the Strategic Leadership Group in Arlington, VA, "…Organizational 'forgetting' drains intellectual capital and squanders the knowledge asset."[1]

The company not only loses the institutional memory along with the old employee, but also must recruit and train a replacement, thereby starting anew rather than building upon the foundation of information acquired over the years by the previous employee. This process can divert valuable resources and consume time that would otherwise be spent on other revenue generating activities. It is far more efficient and less costly to retain talent and keep vital institutional knowledge onboard than to continuously recruit and hire new talent for the same positions. In a competitive, global economy employers who recognize the importance of maintaining a quality, secure workforce with low turnover rates can create value by saving the firm money. They can also create value by properly protecting and training employees, which in turn secures the firm's investment in its personnel.

Many businesses recognize that employees can be the most important factor determining success versus failure. Employees demonstrate value for a company in three key ways. First, employees provide raw intelligence. Employees may have a bachelor's degree, a master's degree, or a professional degree and license, but regardless, degrees present an employer with a wealth of knowledge and skill sets that a company can leverage in the marketplace. Raw intelligence also represents the competitive advantage the business may have over its competition. In a global economy where many businesses struggle to differentiate themselves from competing products and/or services, employees provide intellectual capital that cannot be easily replicated and that translates into value for a company's bottom line.

Secondly, employees provide the employer with valuable institutional knowledge. Institutional knowledge refers to an employee's knowledge of the company, but it also refers to his or her knowledge about the industry in which the company operates. For example, a transportation employee will usually have institutional knowledge—not only about the firm but also about the whole panoply of associated firms that support and sustain operations across the industry and in and around that specific mode of transportation. Why is institutional knowledge so important? According to Harvard Business School, "When employees leave, they take vital knowledge with them. Without a process in place to capture that knowledge and transfer it to their successors, it winds up lost forever. As a result, those who follow them in the job take a longer time to get up to speed, important discoveries and insights disappear and the company's ability to act quickly and intelligently is crippled."[2]

Thirdly, employees generally represent a networking asset to the company. Each employee has his or her own network of contacts that the company relies upon, to some extent, to produce value. For example, a multinational corporation that wants to sell their product in a new market, China for example, often will hire some local talent with appropriate cultural knowledge and regional connections. The new employee can bridge language and cultural barriers and will likely provide contacts in the host country that can be leveraged to increase market share over time. Although it is difficult to quantify the value of contacts, employers in nearly every industry name employee networking or relationship-building capabilities among the most important features of a workforce. Like raw intelligence, a firm should view an employee's set of contacts as a competitive advantage in the market space. Because competitors cannot easily replicate the set of relationships an employee can offer, individual employees give firms a specialized market differentiator. Therefore, companies should try to recruit and retain employees who offer strategically important relationships that can, in turn, be leveraged to grow market share and/or increase revenues.

Wayne Cascio, professor of management in the Business School at the University of Colorado at Denver, found that, "...companies that treat their employees well are also good for investors. Indeed, 80 of 100 companies that made *Fortune's* 2002 and 2003 lists of "100 Best Companies to Work For" avoided layoffs in the prior years... So, the majority of companies on the list strive mightily to provide employment security for their employees."[3] This illustrates the importance of treating employees as assets and communicating that philosophy throughout the corporation, from senior leadership and management to entry-level positions. Cascio's research echoes the sentiment of most employers: It is better to retain talent and cut costs elsewhere, than to layoff talent for short-term cost savings and face having to find and recruit replacement talent in the future.

Companies are beginning to take notice as well. According to a principal at the Boyden Global Executive Search Company, "The truly sophisticated companies are starting to look at a coordinated approach to physical security, information security, and risk management...

People are your company's greatest assets. They provide intellectual capital, raw intelligence, and contacts that, if properly managed, can increase your market share and brand equity. Consequently, a loyal, well-trained workforce may represent your greatest countermeasure to the threats that challenge the global marketplace.

[at one time] companies viewed HR departments as just over-head, until they realized that management of human resources was as critical a business process as any. The same thing will happen with the management of security."[4]

People are a company's greatest assets. They provide intellectual capital, raw intelligence, and contacts that, if properly managed, can increase market share and brand equity. In transportation, they provide the critical linkages that enable a company to achieve operational success and generate profitable business. They represent assets that are unique and therefore have the effect of not being easily replicated by competitors. Consequently, a loyal, well-trained workforce may represent the greatest countermeasure to the threats that challenge the global marketplace.

Trust, But Verify

Trust is the fundamental building block in any business relationship, whether between employer and employee, shipper to forwarder, consolidator to carrier, or retailer to bank. Trust reinforces relationships, which in turn feeds into corporate culture and brand. The transportation industry is diverse and complex, which can provide challenges to creating business relationships. "If there is distrust between workers and managers, or widespread opportunism, then the delegation of authority required in a typical "lean" manufacturing system will lead to instant paralysis."[5] In other words, a business cannot function in a competitive environment if there is distrust between workers and managers. As a Harvard Business School report notes, "By telling the truth and giving people the means to know for themselves, you can build the foundation for a strong relationship...,"[6] In general, policies and procedures that are transparent, that encourage accountability and represent fair and reasonable treatment of employees, can serve to reinforce employer-employee relationships. In order to cultivate human capital, companies should develop and sustain a sense of trust among employees, which can ultimately lead to external business benefits and profitability.

However, it's a fact that there are people in the workplace who will intentionally expose their employers to all kinds of risk and legal jeopardy. While it's important to trust, it is equally important to verify. The reason for this caution has to do with a degree of concern with respect to potential malfeasance, including potential liabilities stemming from misdeeds, theft of goods or corporate secrets, and even workplace violence. According to the Hayes International Retail Theft Survey, one in every twenty-seven employees will be caught stealing from their employers. It is especially important for firms to identify potential internal threats and take steps to mitigate those risks. The ability to protect the innocent while guarding against the potentially troublesome requires a delicate blend of measured security practices and well-implemented personnel policies and procedures.

Both employers and employees should be aware that laws may subject a firm to potential liability in the areas of security, environmental regulations, employment contracts, employment discrimination, sexual harassment, trade secrets, copyright infringement, non-compete clauses, executive compensation agreements, and wrongful termination.[7] Latent liability may also exist related to man-made discontinuous events that can be initiated from inside the transportation industry. Once aware of a potential threat of liability, firms must take affirmative steps to protect themselves, their shareholders (if it is a public company) and the workforce at large.

Securing Human Capital

It's rare—if not unheard of—to hear someone argue against the value of human capital. What does seem to be unanswered, however, is the question of how much is enough when it comes to securing human capital. In general, it is probably safe to say that global firms spend much less effort managing human capital than they do to manage financial capital—but where is the appropriate equilibrium point?

A high percentage of employees spend most or all of their workday at an office or work site. It follows then that a company should dedicate an appropriate, corresponding amount of resources to provide a secure environment for these employees. Many of the best practices mentioned in Chapter 5 regarding securing fixed assets apply here because office and work sites are generally part of a company's fixed assets. Because employees are extremely valuable and difficult to replace, companies should seek to ensure that they have taken reasonable and prudent measures to ensure their security while at work.

Corporations expend a great deal of effort to manage financial capital, and tend to be able to account for the status and security of financial assets, at any given time, in great detail. Is this also true of human capital which most would agree has equal or greater value to most organizations than financial capital? Unfortunately, the answer in most cases is "no." As the firm endeavors to implement TSM and create value, it should be investing significant resources in managing human capital. One component of this investment should include an effective way to account for all of the firm's people (assets) following a catastrophic event. Has the company implemented policies and procedures that would enable a rapid identification of the status of every employee following a disruptive event? This of course is a critical component of business continuity planning, and an essential element in reconstituting a business following a crisis. If employees are missing, does the company have procedures in place to help locate them and provide assistance if required? If employees are temporarily unable to work, are there established policies and procedures that will minimize their absence and assist them in preparing to return to the workforce? All of these initiatives and more are essential components of a comprehensive strategy to secure human capital.

We know from anecdotal evidence that most employees that are precluded from working following a significant disruptive event intend to return to the workforce as soon as possible. Companies can accelerate the return of these assets by establishing policies and procedures that support emergent requirements and provide necessary assistance to facilitate the employees' desire to rejoin the workforce. For example, following the devastation of Hurricane Katrina, and the widespread displacement and temporary disabilities that thousands of employees experienced, The Home Depot announced that they would offer employees whose jobs were affected by the hurricane an opportunity to work at any Home Depot location. This policy benefited employees and built loyalty while at the same time providing The Home Depot with a powerful tool to accelerate business continuity and resilience, and minimize the loss of access to a portion of their workforce.[8]

The Total Security Management approach suggests that companies begin by looking at the four operational enablers in order to determine opportunities for value creation related to human capital.

ENABLER #1: Best Practices Implementation

Today's dynamic security environment coincides with a rapidly changing and highly competitive transportation business climate. In order to remain competitive, firms are forced to deliver goods and services in a timely and error-free manner. Having the right employee base is an essential element of meeting this requirement, and a firm's human capital represents one of its greatest assets. Protecting this asset requires total security processes that create value while not unduly impeding the flow of goods.

One exemplar of best practices with respect to securing human capital is Tropical Shipping, a steamship line based in South Florida, who boasts on their website that, "We've committed to the utmost levels of security to protect our customers and employees and every Tropical Shipping employee is prepared to make security the highest priority." As a result, security awareness training is mandated for all Tropical Shipping employees.[9] This kind of commitment incorporates the fundamental tenets embodied in TSM Pillar Two: *Total Security involves everyone*. In addition to baseline training, companies should also ensure that continual education and safety and security training is available to and taken advantage of by employees and, if appropriate, their families.

Firms should also strive to involve employees at all levels, where feasible, in the security planning process. Utilization of focused process development planning groups can ensure that all reasonable concerns are addressed in terms of physical security, and can provide a company with valuable, field-level input from the employees that are closest to the front lines. The right solution set will vary by location and circumstance, but simple initiatives such as providing employees with a personal emergency preparedness kit (items that an employee would need for at least one day after a disaster or emergency strikes) are minimal investments that a firm can make in each employee that sends a message that they are valued and that the company is concerned with their welfare.

Firms should also strive to involve employees at all levels, where feasible, in the security planning process. Creating a security culture within the firm encourages employees to be vigilant for anomalies or unusual behavior displayed by their peers, customers, or other value chain partners.

ENABLER #2: Situational Awareness

Situational awareness with employees has to do with the need to understand them and their concerns about security. Addressing this need includes all of the physical security and access control matters covered in Chapter 5 as well as the additional security measures outlined throughout this book, including measures to deter theft and terrorism by showing demonstrable vigilance in order to project the image of a hardened target to those that would attempt to exploit or place human capital at risk.

Creating a "security culture" within a firm encourages employees to be vigilant for anomalies or unusual behavior displayed by their peers, customers, or other value chain partners. Such behaviors can be indicative of more serious concerns and potentially

Proper training and exercises will let all internal and external stakeholders, including employees, know that your firm takes security (including employees' personal security) seriously, which can pay off in terms of employee loyalty and reduced turnover.

significant threats, including criminal activity. Employees should have formal mechanisms available for anonymously reporting such concerns and should believe that their reports will be taken seriously and acted upon as required.

ENABLER #3: Training and Exercises

Training and exercises are an important component in the holistic, enterprise-wide approach to securing human capital. By using training and exercises, companies can better prepare employees to respond appropriately to and recover efficiently from discontinuous events, allowing the firm to realize a real return on investment. Well-trained employees have a better chance of defeating a preventable disruptive event by taking preemptive action and training and preparedness has a direct and positive impact on business continuity and resilience. It is not unusual for exercises to produce after action reports and lessons learned that then contribute to business process improvement as well as enhanced security. Proper training and exercises will also let all internal and external stakeholders, including employees, know that the firm takes security (including employees' personal security) seriously, which can pay off in terms of employee loyalty and reduced turnover.

There are many ways to measure investment in human capital, including metrics such as the number of hours of training conducted per employee and relevant certifications attained. The most important metric to measure, however, is the resultant improvement in enterprise resiliency that has been achieved by providing the training opportunities to employees. Resiliency benefits can be evaluated by comparing performance in business continuity exercises before and after training.

The delivery of training information should be instituted through a systematic process using methods that enable an individual or group to learn predetermined knowledge against measurable learning objectives that can be applied to a required standard.[10] There are numerous generally accepted educational and training methodologies currently in use, but maximum return on investment from training professional adults can be achieved using a varied approach including the following three forms of instruction:

- An interactive approach stressing workshops, small group seminars, and extensive tabletop exercises.
- Use of multiple flexible modules to allow for a combination of group and individual activities appropriate to the various attendees.
- Blending of different in-person and distant approaches using classroom training, "webinars," and web-enabled self study.

In addition, various professional associations have established training and/or certification programs that enable transportation corporations or individuals to invest in continuing education for security. Several notable examples include:

- American Society for Industrial Security (ASIS)
  - Certified Protection Professional (CPP)
  - Professional Certified Inspector (PCI)
  - Physical Security Professional (PSP)
- Association of Certified Fraud Examiners
  - Certified Fraud Examiner (CFE)
- APICS
  - Certified in Integrated Resource Management (CIRM)
  - Certified in Production and Inventory Management (CPIM)
- International Warehouse Logistics Association (IWLA)
  - Certified Logistics Professional (CLP)
- Sans Institute
  - Global Information Assurance Certification (GIAC)

ENABLER #4: Outreach

Transportation business can be largely dependent upon personal relationships. Employees at all levels routinely have opportunities to build relationships with internal and external stakeholders and develop a level of mutual confidence that can serve as equity in the event of a significant security event. Following such an event a shareholder, employee, customer, or even an outside third party will be much more likely to assume that a transportation firm did everything that it could have done to avoid the incident if it has previously employed an aggressive and comprehensive outreach strategy. Outreach is therefore an essential element in a value creation strategy. There is value in being seen as a willing partner in implementing security best practices. Companies that place a priority on outreach have realized that close interaction and cooperation with constituents and stakeholders can become a valuable asset over time. For example, Tropical Shipping makes the point on its website that it has become widely recognized as a leader in maritime security, "because of its close and extensive cooperation with responsible authorities, including the Department of Homeland Security."[11]

Applying the TSM Value Creation Model

The next section describes the specific tenets for applying the TSM Value Creation model to the task of securing human capital. It identifies the kinds of issues that a firm should examine internally and externally, as well as the facets that should be examined by analysts, insurers, and regulators.

The Four Value Chain Vs: Visibility, Variability, Velocity, and Vulnerability

Employees are one of the most important assets in a transportation firm because they are directly responsible for the routine application of all corporate procedures, including security. End-to-end visibility can be the hallmark of a secure company, but simply being able to track a shipment is not enough. A company that makes use of that kind of knowledge can better serve its customers and can create value that sets it apart from others. Appropriate supply chain visibility enables security staff and other employees to detect anomalies and allows the firm to resolve minor disruptions before they grow out of control. Variability can often appear to be born out of confusion and therefore generally is undesirable for processes involving securing human capital.

Companies with consistent, appropriate hiring and employee management systems will minimize potential problems caused by employees who may lack the appropriate skills or training for the job they perform.

The velocity with which shipments traverse through the system also is heavily dependent upon human workers, especially when a problem occurs and people need to step in and resolve it. Security practices that increase a firm's ability to know how and where something goes wrong enable employees to provide a better response, which in turn creates value for the firm. In almost every link in most value chain solutions, a shipment is touched by human hands (literally or figuratively), so ensuring that the employees involved have the right level of security and training should, by default, reduce the number of incidents that would slow down the speed of a shipment.

People represent the solution to a majority of vulnerabilities in the transportation system because they are the ones on the scene who can detect an anomaly and then act to correct a situation before it gets out of hand, whether the cause is natural, criminal, or otherwise. Security vulnerability with respect to human capital can be mitigated by adhering to many of the TSM principles that have been presented thus far, including the use of informed hiring practices, the verification and credentialing of all people who have access to logistics facilities, use of access control systems to limit access to those with proper authorization, and the use of technology to detect shipment tampering.

The Four Security Ds: Deter, Detect, Delay, and Dispatch

Deterrence may not be particularly effective against natural threats, but it is fair to say that it has a significant impact on disruptions and losses typically associated with people, including theft, sabotage, vandalism, and terrorism. Criminals are generally rational beings, in that they will choose to strike less well-defended targets rather than appropriately secured ones, an understanding that is the heart of the deterrence approach. As a result, firms with good security practices are generally less appealing targets than those with lax practices. In the same vein, employees who have been trained and tested in security practices tend to be an excellent first line of defense against these types of actions, and offer a better chance for early detection. Detecting a threat or potential disruption falls almost entirely upon the trained perception of human staff, whose job it is to interpret the signs around them and anticipate problems. Employees at all levels of the firm should share the responsibility to detect problems before they occur so that mitigation can prevent loss of business. Being able to delay the threat will enable the firm to devise an appropriate response to the potential problem. Dispatching the appropriate response team in the event of a disruption is paramount to recovering from and surviving the disruption. Properly training response teams and regularly testing and exercising the response plans will ensure that the dispatched people will respond quickly and correctly. All of these elements combined create an environment in which employees are aware of what is expected of them and have the sense that the firm is concerned for their well-being.

The Four Solution Set Cs: Coordination, Cooperation, Consultation, and Collaboration

Although many processes may be aided by technology, at the end of the day they require humans to take action. Hiring the right people with the right skills, and retaining them in order to avoid the loss of institutional knowledge and experience are important for making sure that a firm is effectively working with companies and stakeholders throughout the value chain. Coordination among various value chain partners is obviously necessary for business to function in a secure and efficient manner. In the security arena, coordination is typically most evident in the development and testing of disaster management and business continuity plans that must be coordinated with employees, value chain partners, and other stakeholders in the community.

Cooperation similarly represents a key step in terms of such activities as sharing information with government entities or port authorities and suppliers. It is important that a firm's security staff is in regular contact with others who share its critical security mission because during times of duress these relationships can become critical to the firm's ability to resume business. Security professionals routinely use consultation as a tool to improve the security posture of their company because security is everyone's business, from the distant supplier in a foreign country to the factory worker on the floor, to the loader on the dock, to the security guard at the front gate. It is important that firms cultivate a culture of inclusion and participation where security is concerned in order to encourage cooperation. The sharing of best practices and information in a formal way allows professionals to compare policies and procedures and adapt those that will enhance the overall security posture.

It is important that a firm's security staff is in regular contact with others who share your critical security mission because during times of duress these relationships can become critical to the firm's ability to resume business. Further, the sharing of best practices and information in a formal way allows professionals to compare policies and procedures and adapt those that will enhance overall security posture.

Collaboration also is necessary at all levels to ensure that a mutually supportive and complete security plan is in place throughout the value chain. A business must collaborate with its partners to ensure there are no weak links in the chain. At the same time, a company should collaborate with its employees to devise risk management and response plans that will work. Human capital is critical for all of these aspects and the interpersonal relationships built by employees can make all the difference in a crisis situation.

Conclusion

Arguably, people can be a firm's most valuable asset in the transportation industry. Most firms manage value chains that include a complex mix of competencies, from manufacturing to professional advisory services. The common denominator in all of these value chain networks is people. People also represent a tremendous investment in terms of training, corporate knowledge, industry contacts, and continuity of existing procedures.

A firm should recognize that along with employees come certain inherent risks and responsibilities related to security. It's important to take prudent legal precautions that act in the best interest of all employees by clearly defining acceptable and expected workplace behavior, as well as the corporate policy on a host of other ethical issues. Firms should ensure that all appropriate facility and other security initiatives align with existing best practices for the physical safety of employees (as defined in Chapter 5), as this is an obvious key for physically securing the firm's human capital.

Developing a strategy to secure human capital should involve an enterprise-wide approach that encompasses best practices, effective situational awareness, an aggressive training and exercise regime, and appropriate outreach initiatives. The essential by-product of these activities, and the subsequent return on investment for an organization, is manifested in business continuity benefits that create value by ensuring an organizational structure that cultivates, augments, and retains human capital.

By appropriately training people with respect to security during day-to-day operations a firm can save money by not having to recruit and train new employees, by ensuring the presence of a strong institutional memory and corporate culture, and by cultivating a feeling of loyalty based on the sense that the company cares about the welfare of its employees. Supporting personnel survivability and institutional resiliency is necessarily based upon training. Relevant security training can make it more likely that any looming disruptions might be recognized early and either minimized or even avoided altogether through prompt action.

> Developing a strategy to secure your human capital should involve an enterprise-wide approach that encompasses best practices, effective situational awareness, an aggressive training and exercise regime, and appropriate outreach initiatives. Relevant security training can make it more likely that any looming disruptions might be recognized early and either minimized or even avoided altogether through prompt action, bringing value to the bottom line.

Corporate officers should ask themselves: "Has the firm invested adequate resources in securing human capital?" Investing in such initiatives can ensure that employees, and the firm itself, survive a discontinuous event and resume operations more rapidly. With human capital at the core of the global value chain, industry practitioners must focus on securing this critical asset as part of an overall plan to create value. Once an organization recognizes these challenges and acknowledges the intent to take action, it can make the required changes to get the right resources focused on the right places, at the right time.

Case Study: Proctor & Gamble—Securing Human Capital

Value Chain Security Goals

- Ensure that employees are qualified for the job they are performing, and also have a verified background that has been checked for integrity.
- Use rigorous hiring practices and continuous monitoring to lower the security risk associated with an employee.
- Ensure that new employees receive the training necessary to understand and internalize the corporation's security policies and practices.
- Provide the continuous training and exercises that are necessary to ensure compliance with corporate security and emergency response policies.

Background: Proctor & Gamble

Proctor & Gamble (P&G) is a highly diversified corporation worth $57 billion with global business units operating in over 100 countries. With such a large presence, the security risks are omnipresent. For example, in addition to its 2,500 members of staff, P&G's 37-acre (15-hectare) headquarters site also welcomes more than 800 visitors a day.[12] And with almost 140,000 employees worldwide, P&G is well aware of the value of its human capital as well as the security risk each of those employees represents. Rather than a risk, however, P&G views each employee as a security "opportunity." The company believes security is everyone's business and has been very proactive in its hiring and training processes, in order to ensure that their people understand and support corporate security practices. P&G has embraced the collaborative nature of security in today's global business environment—it is part of their corporate culture and one of their most important metrics.

TSM Solution Set

P&G employee and security-related processes demonstrate the value of human capital and the training that is necessary to leverage each employee's capabilities to secure the organization. Their concern for human capital begins with the initial employment interview process. Potential employees are required, at their own expense, to provide certified background data to the company. New hire training is mandatory and utilizes the P&G intranet for self-training, indoctrination, security contact information, and knowledge of the security network. Continual training and exercises are built into their human resources (HR) processes and P&G tracks each employee's accumulated training as part of the employee's record.

According to Ed Casey, Procter & Gamble's Director of Worldwide Corporate Security, "Our foremost task is protecting our people globally."[13] Casey also supports general information security training and investigations of physical and information security breaches. With P&G's heavy focus on the employee as the locus of security, it is no surprise that Casey reports to HR, which in fact serves as security's point of contact for all personnel. This system works because P&G's first line of security actually relies first and foremost on director-level business managers (called "security champions") who themselves are accountable for all security lapses within their groups. These security champions receive annual training to recognize anomalies and situations that may be indicative of problems such as malicious employee behavior, missing equipment, new vendors, or lax security practices.[14]

Value Creation for Business Processes

- Rigorous hiring practices include verified backgrounds that self-select people with a demonstrated ability to commit to the corporate culture of security from the ground up.
- Making security the responsibility of everyone in the corporation empowers employees to ensure that security practices are followed and to provide feedback to enhance security.
- Training, both at hire and continuously throughout their career, ensures that employees are knowledgeable about company security practices and their roles in the process.
- The use of cross-functional security teams fosters understanding across the corporation and promotes collaboration for process improvements.

Notes

1. Martha Lagace, "Thanks for the (Corporate) Memories," *Harvard Business School Working Knowledge*, 12 May 2003, < http://hbsworkingknowledge .hbs.edu/pubitem.jhtml?id=3456&t=entrepreneurship> (20 April 2006).
2. Martha Lagace, "Thanks for the (Corporate) Memories," *Harvard Business School Working Knowledge*, 12 May 2003, <http://hbsworkingknowledge .hbs.edu/pubitem.jhtml?id=3456&t=entrepreneurship> (20 April 2006).
3. Wayne F. Cascio, "Responsible restructuring: Seeing employees as assets, not costs," *Ivy Business Journal*, November/December 2003, <http://www.iveybusinessjournal.com/view_article.asp?intArticle_ ID=450>, (21 April 2006).
4. Michael Fitzgerald, "All Over the Map," *CSO Online*, June 2003, <http://www.csoonline.com/read/060103/models.html> (29 April 2006).
5. Francis Fukuyama, "Social Capital and Civil Society," *The IMF Conference on Second Generation Reform*, 1 October 1999, <http://imf.org/external/ pubs/ft/seminar/1999/reforms/fukuyama.htm#I> (22 April 2006).

6. Steven Robbins, "The Keys to Building Trust," *Harvard Business School Working Knowledge*, 20 December 2004, <http://hbswk.hbs.edu/item.jhtml?id=4553&t=srobbins> (22 April 2006).
7. Douglas Parker, "Employment & Labor Litigation," *Preston, Gates, Ellis LLP*, 2006, <http://www.prestongates.com/practices/practice.asp?parentPracticeID=141&practiceID=123&showPage=detail> (22 April 2006).
8. John Snow, "Prepared Remarks: The Home Depot," September 16 2005, <http://www.ustreas.gov/press/releases/js2725.htm> (12 April 2006).
9. Tropical Shipping, "Spotlight's On Security," <http://www.tropical.com/External/En/Press/TropicalNews/spotlight_security033006.htm> (28 April 2006).
10. Robin Good, "What Is Your Best Definition of Training?," *Mastermedia.org*, 6 July 2004, <http://www.masternewmedia.org/2004/07/06/what_is_your_best_definition.htm> (29 April 2006).
11. Tropical Shipping, "Spotlight's On Security," <http://www.tropical.com/External/En/Press/TropicalNews/spotlight_security033006.htm> (28 April 2006).
12. ID Tech, "Procter & Gamble invests in the best possible security," *Case Study*, (2006) <http://www.idtech.be/en/references/pg.html> (20 April 2006).
13. Michael Fitzgerald, "All Over the Map," *CSO Magazine on the Web*, June 2003, <http://www.csoonline.com/read/060103/models.html> (20 April 2006).
14. Michael Fitzgerald, "All Over the Map," *CSO Magazine on the Web*, June 2003, <http://www.csoonline.com/read/060103/models.html> (20 April 2006).

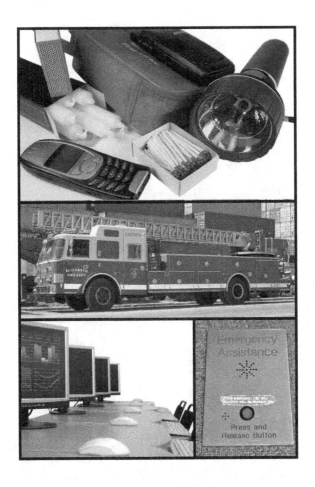

"Clearly, corporate preparedness can mitigate the impact of emergencies on both people and property. Ultimately, preparedness, or the lack of it, can determine the ongoing viability of a firm."

—From the Homepage of New York University's
International Center for Enterprise Preparedness

Chapter Nine

TSM and Business Continuity Planning

Business Continuity Planning (BCP) is a formal process whereby firms analyze risk, evaluate risk mitigation alternatives, and conduct detailed and deliberate planning to ensure a smooth flow of operations in the face of an adverse discontinuous event. It includes (but is not limited to) crisis management considerations, which require that businesses keep up a sustained effort in the development and then subsequent test and evaluation phases of detailed plans for the continuation of core business functions in the face of a significant disruption. Whether the root cause is nature, negligence, or man-made (including terrorism), the effect on the business is the same: the discontinuous event impacts the planned flow of information, goods, and people in ways that affect the value chain and preclude normal operations.

Marsh, Inc., a leading risk management firm, estimates that, on average, companies will face a crisis every four to five years.[1] Disruptions can come in many forms and many can be both planned for and mitigated against. Disaster management plans must be comprehensive and flexible and be able to prepare a company for catastrophic events as well as expected events. According to the chairman and CEO of Marsh's crisis consulting practice, "It's the unexpected crisis that levels the company."[2]

The objective of the firm is to get "back to normal" as quickly as possible, with the understanding that doing so may require close coordination with critical suppliers, rerouting of shipments, the reassigning of resources to certain business areas by drawing them away from non-critical functions, and even, in extreme circumstances, the orderly relocation of some or all of the workforce to alternate sites. However, large-scale BCP is a relatively new endeavor for many firms and one in which, critically, much more work remains to be done by nearly all global transportation entities.

The New York University's International Center for Enterprise Preparedness (InterCEP) observed that:

> "Post September 11th, businesses and other private sector organizations have increasingly acknowledged the need for organization-wide emergency management and business continuity programs. In the United States alone, this need has been validated well beyond the terrorist threat by recent events including devastating hurricanes in America's Southeast, the blackout of the Northeast, tornadoes throughout the Midwest and wildfires in the Southwest.... All of these support the need for an "all hazards" approach to emergency management and business continuity. Clearly, corporate preparedness can mitigate the impact of emergencies on both people and property. Ultimately, preparedness, or the lack of it, can determine the ongoing viability of a firm."[3]

The law of averages dictates that with the ever-expanding number of links in the value chain there is an increasing likelihood of disruption at one of the key nodes. Professor Yossi Sheffi describes what he calls the "high frequency of rare events," when he says, "While the likelihood for any one event that would have an impact on any one facility or supplier is small, the collective chance that some part of the supply chain will face some type of disruption is high."[4] It is easy to see this potential, especially when it is multiplied by various threats throughout the broad and deep and interconnected network of transportation and related businesses that interact to deliver goods from the field to the end user. At the same time, as a recent study commissioned by AT&T concluded, "Ignoring risk does not make it go away. However, firms can manage and live with risk...effective BCP is more than simply keeping critical data in more than one spot; it is a structured and formal process that identifies, manages, and reduces all forms and types of supply chain risks."[5]

Because business continuity planning is critical to ensuring the survivability of a firm and its relative stability during significant discontinuous events, it creates value by contributing to the firm's long-term viability. When properly demonstrated and used as a market-place differentiator, BCP also helps to build critical consumer confidence that the firm will continue to operate in the face of adverse events, which in turn can pay dividends in terms of customer loyalty. For example, The Home Depot has taken the initiative within its global value chain networks to ensure that it sources merchandise from numerous vendors worldwide, thus eliminating single points of failure. With 600 factories in approximately 35 countries The Home Depot has demonstrated to analysts and shareholders that it is not dependent on any single vendor or country.[6] It is this kind of focus on real-world survivability and ensuring measurable, defined value that explains why business continuity planning serves as one of the cornerstones of the Total Security Management approach. Indeed, it is such a significant part of TSM that it is singled out as the basis for the final of the Five Pillars of TSM: *Total Security requires resiliency and business continuity planning as essential business functions.*

According to Michael Heath, Vice President of Sales for AT&T, "a static business continuity plan will not protect a

Key Aspects of a BCP Strategy

- Addressing measures to ensure the viability of those critical functions that enable the firm to continue operations.
- Outlining decision processes for implementing BCP measures.
- Identifying procedures for employee communications and issuing alerts/updates.
- Enabling a 'head-count' to locate all employees, establishing their location and ability to perform critical tasks.
- Including activation of a public relations crisis management team.
- Beginning the process of assessing and acquiring needed resources for the period of disruption.

company from disruptions... a robust plan that ensures continuation of services during disruptions requires a supply chain component and regular testing."[7] This is why BCP plans need to be developed, tested, and built into the business process long before they are needed, and then regularly exercised to ensure all participants can play their respective roles without hesitation. Taking steps such as properly designing and implementing a firm's cargo tracking system, forging the right relationships with value chain partners, and monitoring the security of personnel and facilities will immediately help arm the firm with the information it needs to proceed. Once it has a plan in place, the company is able to react with confidence and purpose when a current or impending disruption occurs in its transportation network.

A firm's response plans should cover a variety of possible incidents, ranging from predictable acts of nature, to accidents, sabotage, or terrorism. It should also take non-physical incidents into account, such as the kind of political instability that can disrupt a value chain—striking workers in a distant port or critical infrastructure capacity limitations. The company should plan ahead for flexible sourcing, perhaps having multiple suppliers in different locations, or local supply sources for smaller quantities to get through a crisis. The firm should have already mapped out and made arrangements to call on alternative ports in case they're impeded from using the scheduled ones.

Additionally, there should be a plan in place to isolate and deal with a shipment when sensors indicate possible tampering. For example, in such cases a process should be in place for alerting proper authorities and determining the requisite level of response.

Successful development of the incident or crisis management plans will include engaging and training all personnel at all levels. This includes the partners in the value chain, for they too should have contingency plans that match and feed into the firm's plans. It also follows then that firms should build this requirement into solicitations and use this as part of the criteria for performing due diligence when evaluating value chain partners. It is important that the plans are well known and understood, that training has been given and updated regularly, and that mock exercises have been undertaken to ensure that the response will be second nature and will not require significant ramp-up time. In this manner, BCP again exemplifies the TSM paradigm, because it involves all elements of the value chain, including considerations on everything from suppliers, carriers, terminal and port operators, to the activities of various governments and international regulatory bodies. These latter two categories play an especially significant role because the firm's response alternatives will likely be dictated at least in part by regulatory and international movement-of-goods decisions over which the firm has little or no control, such as temporary suspension of certain modes of transit, closure of ports, quarantine of large areas, or heightened cargo and related inspections requirements. A comprehensive BCP addresses these concerns by aggressively mitigating single points of failure in the value chain and determining practicable work-around solutions for a variety of complex potential disruptions.

As part of a firm's business continuity plan it should also create a documented recovery plan with defined steps for reestablishing critical operations following a disruptive event. This plan needs to include an internal protocol for reinstating or replacing lines of control and recovering and restarting the information systems necessary to conduct business. During 9/11 several companies were faced with large loss of life and institutional knowledge. They lacked the planning and procedures to enable them to rapidly reinstate operations and were without the leadership necessary to immediately get back to business. Appropriate exercises integrated in the value chain need to be regularly conducted to ensure that the personnel can overcome the shock of the incident and impose a return to a degree of normalcy. This type of crisis recovery plan will probably involve both the private and the public sector since these types of incidents, terrorist attacks, and significant acts of nature on the scale of the hurricanes and tsunamis the global economy recently endured, generally involve a mix of both governmental and private business response and recovery operations.

FIGURE 9.1   Key aspects of a BCP strategy

Approaching BCP

Many of the analytical tools used in business continuity planning are similar to those used to deal with risks to day-to-day
operations through the risk management process. As with risk
management, the key to BCP success is to properly identify
which critical business functions must be continued in the
face of adversity. Accordingly, the BCP process typically begins
where the firm's risk assessment leaves off, with the first order
of business being to review and tailor the defined list of critical
functions and business processes to determine in what way the
firm's operations and value chain are most likely to be affected
by various events. Armed with such knowledge, a business
continuity planning team can determine where to focus efforts
and which specific critical functions must be maintained and

at what relative priority. It is here that they should also look for linkages and critical interdependencies, specifically those that they can make more resilient in order to create broader gains across all manner of preparedness risks.

As explained in Chapter 4, risk management involves the trade off between upfront security costs versus the potential—and potentially catastrophic—costs of a significant discontinuous event, by empowering senior decision makers with a deliberate approach to risk reduction and/or mitigation. The BCP focus, conversely, is on ensuring the firm has properly planned for emergency operations that will enable it to function in the face of an event that already has transpired. In other words, whereas risk management seeks solutions for eliminating or minimizing the risks that are presented in a variety of daily activities in order to minimize the likelihood of disruptions, BCP accepts that disruptive events will occur—and it plans for them accordingly. BCP deliberately seeks out ways to ensure an orderly continuation of the most important business functions throughout the period of disruption.

Indeed, according to a recent report from the Eli Broad Business School at Michigan State University, "While companies may not be able to drive out the fear factor [created by terrorism], a secure value chain, with a system-wide plan to detect, prevent, respond, and recover from value chain disruptions, is a powerful defensive weapon in the hands of business."[8]

BCP Focus Areas

The BCP provides a flexible but codified structure which guides management's decisions and delineates the details of the appropriate business and technical response. The plan includes high-level business continuity determinations such as the goals of the firm during periods of crisis and the relative importance and rank-ordering of various corporate functions. It also includes detailed specifics about who to contact and how to do so, as well as detailed descriptions of anticipated

The BCP process must balance the need to invest resources in redundant communications against the equally important need to ensure the continuation of the processes that the communications exist to support.

BCP Plan Essential Information

Here are some must-haves for a BCP plan:

- Contact list for notification and activation of BCP plan.
- Overview of recovery strategy and order of resumption of critical functions.
- Detailed resource requirements for set periods of time (one week, one month, longer).
- Contact list for critical value chain partners.
- Relocation and off-site workflow plans.
- Data restoration and remote access plans and procedures.
- Delineated legal chain of succession in the event of loss of life or communications with top-level management.

resource requirements for recovery efforts. But more than anything else the BCP process is about putting relative values in a prioritized order to guide the allocation of resources and efforts in time of crisis.

For example, communications with suppliers and vendors is paramount for any firm, and may drive the BCP team to determine that redundant communications networks are necessary expenditures in terms of ensuring the firm's ultimate survival. At the same time, working communications networks are useless without access to the useful information that they are intended to convey—such as the information contained in myriad corporate databases and shipment processing logs. The BCP process must balance the need to invest resources in redundant communications against the equally important need to ensure the continuation of the processes that the communications exist to support. This represents another case in which investment in RFID and GPS technology that supports end-to-end visibility could pay an additional dividend to the firm. Using the real-time information that these systems provide will give a firm an advantage by enabling it to make quick decisions about rerouting, diversion, or even replacing a potentially delayed or disrupted shipment.

The firm also must continue processing bills of lading to ensure that documentation does not become the pacing item for freight movement. Transiting through choke points (such as the Panama Canal) and entry into certain preferred ports (such as U.S. ports that require advance notice of the vessels' contents) are critical and time-sensitive events that require synchronization throughout the value chain. Transportation firms must be able to ensure that all required functions can be carried out on-time and across the globe, regardless of local or regional disruptions. Therefore, a resilient capability for tracking in-transit shipments may also be critical for continuity of operations, and back-up sites or plans for shifting work requirements to alternative locations must address the concerns of data preservation and remote systems accessibility.

Key Emergency Preparedness Terms*

Mitigation
Implementing actions that actually eliminate or reduce the occurrence of a disaster. It also includes long-term activities that reduce the effects of unavoidable disasters.

Preparedness
Activities necessary to the extent that mitigation measures have not or cannot prevent disasters. Such activities include developing plans to save lives and minimize damage as well as to enhance disaster response operations.

Response
Activities designed to provide emergency assistance for casualties and reduce the probability of secondary damage.

Recovery
Activities that continue until all systems and processes return to normal. Short-term recovery includes stabilizing of minimal operating standards, while long-term recovery may take years.

*Adapted from FEMA's 'Emergency Program Manager' Independent Study Course available at http://training.fema.gov/EMIWeb/IS/is1lst.asp

Common Pitfalls in BCP

- Failure to incorporate all levels of staff in the planning process and ensure enterprise-wide awareness of BCP plans.
- Delegation of planning to subordinate offices through generalized templates without ensuring full compliance with the intent and scope of the required planning.
- Over-emphasis on computer systems and data recovery without focus on the human capital aspects of BCP and considerations for physical safety and security.
- Lack of evacuation and physical survivability planning.
- Failure to test plans through realistic exercises.
- Failure to update plans on regular basis.

BCP is really about business process prioritization and focusing response and recovery efforts where they will best support the continuation of necessary business functions. One common approach is to devise categories for the relative importance of critical functions, and a corresponding timeline for restoration. Each of these categories can be further divided into subsections that specifically delineate their relative criticality, while also defining the subtasks and associated recovery timelines. Backup recovery sites and data centers are an excellent example of the best practices in this field, with many being designed so that critical operations can resume as soon as the people arrive at the alternate site. However, for transportation firms, the scope of the focus must be on more than just data systems, for external communications and an understanding of the event's impact on the global transportation network will be critical to making appropriate business continuity decisions.

Representative Business Function Categories:

Category 1 (Immediate Recovery)
Functions of primary criticality that must be addressed first to ensure continuous operation or, if impeded, immediate resumption, include:

- Movement of just-in-time goods throughout the value chain.
- Communications and tracking systems that enable discussions with value chain partners to determine resiliency analysis/awareness and early detection of emergent problems and concerns.
- Outreach to critical customers and stakeholders.
- Public relations initiatives related to the incident and recovery efforts.
- Legal aspects of the incident and any associated liability and recovery issues.
- Government relations, as applicable.

Category 2 (Initial Workarounds)
Important but less economically significant or time-constrained functions that may be temporarily suspended, or whose resumption will be delayed by up to one week, include:

- Engineering and facilities maintenance concerns.
- Accounting functions not related to the incident.
- Collection of accounts payable from certain trusted partners.
- Other non-critical backroom processing of payments and related information.

Category 3 (Temporarily Suspended)
Other business functions of lesser significance that can be suspended indefinitely or until the situation returns to normal, include:

- Advertising and marketing matters (excluding initiatives related to the incident and recovery efforts).
- Planned changes to non-critical human resources policies and procedures.
- Certain capital expenditures on equipment that is serviceable but nearing the end of operational utility.
- Personnel evaluations.
- Product development.

Evacuation and Asset Flow Considerations

Another important area where optimal business practices and business continuity management overlap is in the area of facilities design for the accommodation of the rapid movement of people and goods. The benefits for routine transactions are obvious, for more rapid processing and shorter delays reduce costs to the owner/operator of the infrastructure, and also pleases clients, which is useful in building brand loyalty. The business continuity aspects of proper flow design are potentially even more significant during crisis events, however, for during an evacuation such design can mean the difference between life and death for a company's most critical resource—its people.

One example of a solution provider that is using best practices to help firms deal with this challenge is Regal Decision Systems, Inc., a Maryland-based software services and technology firm specializing in operations research, simulation software, and process flow management. Regal's primary specialty is to provide firms with decision support tools that facilitate decisions related to evacuation planning, staffing, facility designs, trade corridors, traffic management, and emergency preparedness. Their approach is to capture actual data on the movement of intermodal commercial transportation assets, vehicles, passengers, and pedestrian traffic under various conditions and then feed this data into their proprietary modeling and simulation platforms, in order to create highly accurate three-dimensional representations of the flow of assets. These models apply to both normal process flows (such as staffing of baggage screening or inspection of international commercial traffic) and the fluctuations that accompany various crisis events (such as personnel rushing out while first responders rush in, as well as the capacity impacts of a disruption over time). In this way, events and operations are modeled to determine optimal asset flows and best manage traffic demands. For example, three of Regal's products have special importance to transportation firms. The first is their Seaport Capacity Model, which is a discrete-event simulation model that can be used to evaluate operational changes, design changes, and policy changes at a seaport. The heart of the system consists of a graphical user interface (GUI) that allows the user to adjust various properties of the assets, add data, change configurations, and run simulation experiments. The model also provides a two-dimensional visualization of the operations, a statistical output in the form of tables and graphs, and can be used to evaluate the operational impact of various new security initiatives. It can also evaluate the impact of changes to the transportation network that services access to the port.

Regal's second noteworthy product was designed to help stakeholders with ship rerouting decisions in the event that an emergency closes down a given port. Using Regal's "Operations Restore" decision aid, stakeholders can allocate each cargo type a scaled priority rating, in order to determine the efficient reassignment of the appropriate vessels to the right ports, which can automatically cross-sort the assigned ports for the appropriate loading/off-loading of equipment. The model also looks beyond the port to the adjacent traffic network to determine if the roadways can handle the extra truck traffic produced by the port.

Finally, the Evacuation Guidance System by Regal is a software system designed to provide real-time intelligent guidance to the evacuation of employees from a facility where security or structural integrity has been compromised. Its objective is to provide an intelligent, dynamic evacuation route for all occupants by taking into account the location and quantity of any air contaminants or fires and optimizing an evacuation strategy based on time, efficient use of stairs and doors, and limited cross-contamination. Guidance is visually displayed to the evacuees via a series of light segments that provide direction and eliminate confusion during stressful evacuations. Use of such tools can dramatically increase the survival rate of employees if the worst should occur.

Assessing Value Chain Partners

The preparedness and resiliency of value chain partners are key considerations for BCP. The appropriate level of scrutiny will be determined by the nature of the relationship, the size of potential impact if the partner's activities are affected, and their current security posture in terms of physical, communications, and

Top 10 Errors in Emergency Response Plans*

1. Lack of upper management support.
2. Lack of employee buy-in.
3. Inadequate employee training on the plan.
4. Insufficient practice of the plan.
5. Lack of designated leaders during emergencies.
6. Failure to update the plan to reflect recent changes.
7. No means to communicate alerts to employees.
8. OSHA regulations not accounted for in the plan.
9. No procedures for shutting down critical or sensitive equipment.
10. Employees not told what specific actions to take in the event of an emergency.

*Adapted from South Carolina's Workplace Security Guide, February 2003

personnel security. Also, it should be noted that assessing these partners could be time-consuming at the beginning, because all existing critical partners must be evaluated, but over time requires less effort. This is because businesses need only conduct periodic reviews of existing partners and full assessments only for new partners. However, businesses must make BCP a condition of doing business. This is an opportunity to collaborate, share best practices and include partners in BCP planning and exercises—as well as demonstrate to value chain partners that the company places a real bottom line business value on proper BCP.

> You can demonstrate to your partners that you place a real bottom line business value on proper BCP by making BCP a condition of doing business. This is an opportunity for you to collaborate, share best practices and include your partners in your BCP planning and exercises.

Other issues to review might include their previous experiences and response to discontinuous events and the empowerment of their management in taking rapid corrective actions when facing a problem. Similarly, the financial health of the partner firm should be considered, for unhealthy firms tend to cut back on otherwise routine inspections and good security and safety processes.[9] Such information can often be obtained from media review of stories involving key value chain partners.

Another focus area is the partner's communications with the owners/operators of key transportation nodes, such as terminal/port operators and the carriers that they rely upon. Finally, it's useful to determine whether or not they have appropriate procedures in place to screen their value chain partners, and their past record in following through on implementing decisions guided by such reviews—or if they simply offer compliance waivers when problems are found. Keep in mind that choices for ports/terminal operators and, even to some extent, carriers are generally limited. Firms often cannot exert significant market pressures to ensure that these partners have robust BCPs (for example, they can't simply take their business down the street to the next port or railroad). Consequently BCP can and should address crisis mitigation and options for whenever one of these critical links might be disrupted.

BCP Implementation

The final step in the BCP process is developing a clearly defined process for implementation, which is essential because orderly implementation has a significant impact on sustaining and/or resuming critical operations. Many firms create "emergency response teams" from the ranks of existing corporate leadership, typically requiring four teams to deal with such events, although the size of the teams may change as events unfold and the magnitude of the disruption becomes clearer. Each of these teams has significant responsibilities which impact upon each other but should not directly overlap. The four most common teams used are:

· Notification Team
· Emergency Coordination Team
· Recovery Team
· Response Team

Using this model, notification of a potential disruption is passed initially to the Emergency Coordination Team, which is the primary point of contact for employees or others to report potential threats or disruptions. The Emergency Coordination Team relays the information to the Notification Team, which is usually comprised of senior-most management from each operating area, in order to ensure that BCP response measures are not unduly invoked. These two teams collectively determine the likelihood and significance of the potential event's effect on operations. Meanwhile, the Emergency Coordination Team continues to monitor events while the Notification Team reports to the firm's senior-most leader (typically the Chairman or the CEO) in order to receive a decision about enacting BCP measures. If enacted, the Notification Team notifies the Response Team and the Recovery Team, who then begin to enact the appropriate corrective and protective measures, including relocation of personnel, outreach to value chain partners, and rerouting of goods in transit, as applicable.

During the time leading up to and at the onset of a crisis event, the Emergency Coordination Team manages emergency response activities by collecting information from the Response and Recovery Teams, such as status reports, facilities, and infrastructure damage assessments, and any known personnel injuries

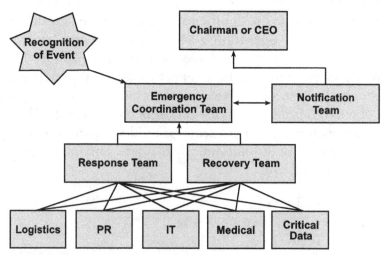

FIGURE 9.2   BCP emergency response teams

or other health and safety considerations. If necessary it can harness additional resources to establish sub-teams for addressing logistics, public relations, medical, information technology, and critical data preservation and recovery. The Response Team's main focus is on continuity of operations in the immediate moment, while the Recovery Team concerns itself with preparations for management of prolonged dislocation and longer-term normalization of business functions.

Conclusion

Business continuity planning is an essential pillar in the implementation of Total Security Management. BCP enables companies to more appropriately evaluate operational processes and to prepare for discontinuous events in a rational and deliberate manner.

Perhaps most importantly, business continuity planning can create value for the firm. This value may not necessarily be easily measured directly in dollars, but is significant and should not be ignored. The created value (which is realized as the BCP process is implemented), exists in the fact that BCP minimizes a firm's downside risk. In this sense, BCP is like any other form of insurance against risk: it minimizes overall risk to acceptable levels by diminishing the likelihood of catastrophic loss, costing a little each day but there when it is needed. Seen in such a light, BCP becomes not an economic drag, but rather an efficient means to mitigate operational risk, and therefore to ensuring the longer-term viability of the firm.

Two especially significant BCP considerations are a) the need for predetermining a priority ordering of the restoration of select critical operations and, b) the need to invest in appropriate design processes in order to "build in" protections that help to ensure the safety and security of staff (including potential evacuation and relocation considerations). As has been discussed throughout this book, the need to exercise and train staff in business continuity plans remains paramount, as does the need to reach out to various neighbors and other interested stakeholders. In the end, given the growing risk posed by the likelihood of future discontinuous events, BCP may well represent the Total Security Management investment that does the most to ensure the survivability of the firm in the face of significant challenges.

Notes

1. Marsh & McLennan Companies, "Risk Reports," (2006), <http://www.marsh.com/MarshPortal/PortalMain?PID=AppDLNoRM&t=11459739 95763&4=AppDLNoRM&3=Overview&2=CrisisManagement&1=Risk Issues> (25 April 2006).
2. Eric Krell, "7 Corporate Red Flags," Business Finance Mag.com on the Web, August 2002, <http://www.businessfinancemag.com/magazine/archives/article.html?articleID=13891&pg=7> (20 April 2006).
3. New York University International Center for Enterprise Preparedness, "About Us," (2006) <http://www.nyu.edu/intercep/> (14 April 2006).
4. Yossi Sheffi, The Resilient Enterprise (Massachusetts: The MIT Press, 2005), 26.
5. George A. Zsidisin, Ph.D. and Gary L. Ragatz, Ph.D. and Steven A. Melnyk, Ph.D., "Effective Practices in Business Continuity Planning for Purchasing and Supply Management," 21 July 2003, <http://www.bus.msu.edu/msc/documents/AT&T%20full%20paper.pdf> (9 April 2006). This report was produced by Michigan State University from one of five grants totaling $250,000 that the AT&T Foundation made to U.S.-based universities to identify a more formalized approach to business continuity planning that would help organizations identify and quantify risks and then implement procedures, strategies, and tactics aimed at ensuring business continuity, thus demonstrating their recognition of the need for businesses to deal with these complex challenges.
6. Securities and Exchange Commission, "Form 10-K for The Home Depot, Inc.," 30 January 2005, <http://www.sec.gov/Archives/edgar/data/354950/000104746906004211/0001047469-06-004211-index.htm> (28 April 2006).
7. George A. Zsidisin, Ph.D. and Gary L. Ragatz, Ph.D. and Steven A. Melnyk, Ph.D., "Effective Practices in Business Continuity Planning for Purchasing and Supply Management," 21 July 2003, <http://www.bus.msu.edu/msc/documents/AT&T%20full%20paper.pdf> (9 April 2006).
8. Diane Closs, "Security Practices Safeguard the Supply Chain," Michigan State University BROAD Business on the Web, 2005, <http://bus.msu.edu/alumni/publications/broadbusiness/05/reality2.cfm> (9 April 2006).
9. George A. Zsidisin, Ph.D. and Gary L. Ragatz, Ph.D. and Steven A. Melnyk, Ph.D., "Effective Practices in Business Continuity Planning for Purchasing and Supply Management," 21 July 2003, <http://www.bus.msu.edu/msc/documents/AT&T%20full%20paper.pdf> (10 April 2006)

"This is not the end. It is not even the beginning of the end. But it is, perhaps, the end of the beginning."

—Sir Winston Churchill,
British Prime Minister during World War II

Chapter Ten

The End of the Beginning

The global transportation industry prioritizes, orders, ships, tracks, and delivers several trillion dollars worth of commercial goods every year using international maritime, railroad, aviation, and trucking infrastructure. This is a complex system, which includes thousands of ports and airports, millions of miles of highway and rail lines, tens of thousands of bridges and tunnels and millions of transportation industry workers. On September 11th, 2001, the world learned just how vulnerable this transportation system had become. When 19 terrorists, posing as regular travelers, turned four commercial passenger jets into lethal missiles—they exploited a significant gap in the security of the global transportation network and were able to "weaponize our transportation system."[1] It seems almost inevitable that future terrorists will attempt to use our highly open, global transportation system against us. These threats could surface in a variety of forms, including truck bombs, like those used in the 1993 World Trade Center and 1996 Oklahoma City bombings; or attacks on ships using small rafts filled with explosives, as with the USS COLE in 2000 or the French oil tanker M/V LIMBURG in 2002. It could be backpack bombs, such as the Madrid and London train attacks of 2003 and 2005, respectively, or even the highly-publicized threat of chemical, biological, radiological, nuclear or high explosive devices smuggled in through an ordinary shipping container. Those responsible for securing critical transportation infrastructure have done a lot of work over the last four years, but the goal of retrofitting security infrastructure on an industry that, with the exception of the aviation segment, had very little to begin with, is a daunting and expensive undertaking.

And yet terrorism is far from the only threat to global transportation networks that justifies the use of better and more systematic value-added security processes. In fact, this book has consciously steered away from making terrorism the primary focus because, even without terrorism, the confluence of the three Change Agents—*globalization, interdependent infrastructure and economic interdependencies*, and *discontinuous events*—has resulted in more disruptions taking place in recent decades, with larger cascading impacts, than ever before. Many of these disruptions happened independent of any terrorist activity. This trend is likely to accelerate as we continue into the future. In response, transportation industry stakeholders must take appropriate steps to ensure all relevant security and resiliency considerations for critical business functions are addressed. It is no exaggeration to say that the economic health of the global free market economy depends upon the market-driven decisions and collective security actions taken by the transportation industry at large.

> It is no exaggeration to say that the economic health of the global free market economy depends upon the market driven decisions and collective security actions by the transportation industry at large. With regard to transportation, it is the entire industry that is at risk collectively as opposed to individual companies per se. The fact is that there is tremendous collective risk that requires collective action.

Unfortunately, with respect to transportation security, it seems that this situation of mutual dependence and shared responsibility has manifested itself in what economists call the "tragedy of the commons." This phrase refers to how certain commonly beneficial actions will not be undertaken because the expenses borne by some will benefit others—whether or not they also pay their share. These so-called "free-riders" then gain an economic advantage that they can transform into lower prices or excess profits.[2] With regard to transportation, it is the entire industry that is at risk, collectively, as opposed to individual companies *per se*. Without a model for creating value through security, each firm may decide that it does not have adequate incentives to secure itself unless its industry partners follow suit for two very clear reasons. First, costs of business will increase relative to those of their competitors, which is bad for business, and second, even if they are better protected individually, their industry is not, meaning that the adverse event is still likely to occur. This reality counteracts the fact that there is tremendous collective risk that requires collective action.

The Total Security Management Solution

The Total Security Management approach addresses this concern by allowing the firms that embrace this methodology to actually create value, which in turn gives them the edge over the "free-riders," who are left behind. At the same time, TSM offers a concise, deliberate means of assessing and analyzing a firm's overall security and resiliency. TSM takes account of the firm's internal security practices and then compares them against industry best practices. At the same time, TSM also evaluates the firm's ability to leverage its awareness and control of its key resources (fixed assets, assets in transit, brand/goodwill, and human capital) in order to implement business and security improvements. Next, TSM looks at both the implementing firm and its value chain partners to verify compliance with regular training, exercises, and outreach to interested parties, including regulators, law enforcement, and neighbors. Finally, TSM examines the firm's preparedness in terms of business continuity planning.

The goal for TSM implementation within the transportation industry is the development of corporate policies that focus on all hazards security threats that have the potential to disrupt or destroy transportation infrastructure, routes, or the transactional data that drives commerce.

The goal for TSM implementation within the transportation industry is the development of corporate policies and procedures that focus on all-hazards security threats that have the potential to disrupt or destroy transportation infrastructure, routes, or the transactional data that drives commerce. In order to optimize return on investment, the focus for TSM implementation must be on the firm's key assets mentioned above. The detailed analysis that results from TSM initiatives can support internal decision making where proper resource allocation across enterprise activities is concerned. It can also be useful for external customers and partners, who will benefit from working with a firm that is better positioned to meet the reality of today's challenges. In addition, external stakeholders such as industry analysts, investors, regulators, and underwriters can benefit from a systematic and deliberate approach to assessing long term value and resiliency, in turn offering the firm economic rewards for having implemented appropriate security solutions.

Companies that become proponents of TSM will be market leaders. They will collaborate, share best practices, and mentor their partners. They will demonstrate long-term value and resiliency, and will reap the economic rewards for having implemented appropriate security solutions.

The reason that firms should be willing to invest to protect their value chains is that security breaches, work stoppages, and business interruptions caused by both natural and man-made events all have the ability to cause significant economic loss and, in some cases, a loss of life. What's at stake for the global transportation industry is the potential for disruptive events to directly or indirectly impact their core business. As noted by Deloitte Research, "The secure economy is characterized by a fundamental shift in the way security is viewed by companies and governments: while once mostly signifying the physical protection of assets and people, the concept of security has taken on a broader meaning. It now stands for sustainability and the ability to make rapid adjustments to the business, to enforce compliance, and to absorb unforeseen costs—all essential components of managing a business."[3]

Companies that become proponents of TSM will be market leaders. They will collaborate, share best practices, and mentor their partners. They will demonstrate the value of TSM by example. They can establish the direct rewards of being compliant with TSM practices by rewarding the other companies within their value chain. This reward might be in the form of preferred vendor agreements, guaranteed percentage of business within a market segment, or even just public recognition. In fact, using public relations as a tool to showcase TSM can reinforce consumer confidence, attract new customers, and inspire customer loyalty.

Sound TSM practices require frequent interaction, on many levels, among all partners. Communications and information sharing minimizes uncertainty, promotes collaborative solutions, and most of all, enables firms to predict with some certainty how their value chain will respond to discontinuous events. Collaborative activities can provide a foundation for building the trusted relationships needed to survive in the global economy.

Supporting and encouraging value chain partners to adopt TSM can have a multitude of benefits that will ultimately strengthen the entire global transportation network. Put simply, secure partners help to ensure security. Secure partners tend to be the more resilient partners and therefore more stable for strategic business reasons. In the past, firms have repeatedly demonstrated that success is contagious. Partners will tend to be loyal to those business relationships that prove to be equitable and profitable in the long run.

The transportation industry at large can benefit from the cumulative effect and market forces that multiple simultaneous security initiatives, guided by a standard TSM implementation methodology, can create. TSM can help shepherd the industry into an era of uniform application of security initiatives, using fundamental and standard principles designed to support the concept of managing security as a core business function, and restructuring security practices to create value.

With the emphasis on best practices, situational awareness, training and exercises, and outreach, the TSM approach brings order and structure to the efforts of business process reengineering related to security. By identifying opportunities to enhance efficiency while reducing risk, and focusing resources on TSM implementation efforts in these areas, firms have the opportunity to synchronize security initiatives and foster better value chain relationships.

Looming Regulatory Response?

Government reaction to a perceived marketplace failure, such as that facing global trade security, can come in the form of regulations that force compliance. One significant factor in the decision about whether or not to pursue this industry-wide approach to TSM is the recognition that, even as this book goes to press, there is pending legislation that would require the U.S. Department of Homeland Security to create a port security strategy delineating minimum security standards affecting most or all shippers. As described by Logistics Management, "The SAFE Port Act seeks $800 million per year for five years to track cargo containers en route, screen port workers, and develop a worst-case scenario plan for resuming

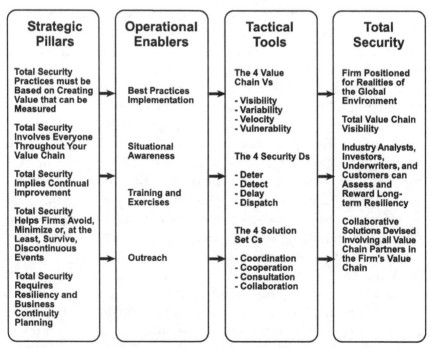

FIGURE 10.1   The TSM approach

operations in the case of a terrorist attack, among other things. The bill also directs Homeland Security Secretary Michael Chertoff to create a port security strategy with minimum standards for checking cargo containers and a plan for getting ports operational again in the case of an attack."[4] If achievable, such legislative initiatives may create a baseline of security practices that represents the bare minimum required for something approaching secure commerce in the U.S., but if the history of government regulation is any guide then it is likely to do so with sub-optimal efficiency. The reality here is that governments are often forced to use a one-size-fits-all approach, and in general, tend to be too inflexible to optimize solutions through regulation for a dynamic, and rapidly and continuously evolving industry.

Government reaction may come in the form of regulations that force compliance. Such initiatives may bring about a baseline of security practices that represents the bare minimum required for something approaching safe commerce, but if the history of government regulation is any guide then it is likely to do so with sub-optimal efficiency. A better response than additional government regulation would be one driven by free market economic imperatives. TSM encourages such a solution.

Collective Self-Regulation

Fortunately, where there is shared risk there is also potential for shared gains. The clearest path toward optimizing transportation security solutions would be one driven by free market economic imperatives. This solution would necessarily be "of the transportation industry, by the transportation industry, for the transportation industry." The "carrot", of course, must exist in reward for firms who implement coordinated, verifiable security best practices in order to create value. Although the specific solution set for any particular company will be dictated by its unique circumstances, this book has included multiple examples where TSM-compliant process improvements have already increased security and provided additional value creation opportunities—especially in the areas of securing fixed assets, assets in transit, human capital, and brand equity and goodwill. Indeed, value creation is the genesis for all five pillars of TSM.

The key to creating the requisite market-driven solutions for transportation security will depend upon the synchronization of three focus areas:

1. Recognition of the need for change by key industry players.
2. Acceptance of a common framework for mitigating value chain security risks.
3. Concerted self-regulation and assured implementation of best practices.

Recent signs have emerged, however tentative, that the private sector is beginning to think more seriously about implementing such a self-regulatory approach. For example, the former Director of Supply Chain Assets Protection for discount goods retailer Target Corporation concluded in an interview, "The trade community has come to realize that we need a system— a trust-building measure—that will hold each stakeholder accountable for securing their particular link in the chain."[5] A July 2006 report from Stanford University similarly concludes, "We strongly urge companies not to consider security investments as a financial burden, but rather as investments that can have business justification, that can result in operational improvements, and that ultimately may promote cost reduction, higher revenue, and growth."[6] Finally, even as this book is being written, New York University's International Center for Enterprise Preparedness (InterCEP), which bills itself as the world's first major academic center dedicated to private sector crisis management and business continuity, is preparing to host an "International Conference on Emergency Preparedness Standardization." InterCEP says the conference is being held in response to the need for an international standard for emergency preparedness and to promote the development of a framework to be published by the International Organization on Standardization (ISO), whose work was instrumental in the eventual large-scale adoption of many of Dr. Deming's principles of Total Quality Management (TQM).

In terms of the second focus area—the development of an appropriate framework for mitigating value chain security risks—Total Security Management was specifically designed to serve as the basis for a common approach to verifiable corporate best practices in security, throughout the global transportation network.

Finally, the third focus area, that of the concerted self-implementation and self-regulation of best practices, will only come about once the private sector agrees that security matters. Much in the same way that the business world was forced to reconcile the relevance of quality in the past, firms must decide if security is important enough to be managed as a core business function. Once this determination is made, firms can initiate an enterprise approach to security that includes identification of opportunities for return on investment in a host of traditional and non-traditional initiatives. In today's complex and interdependent economy, transportation firms around the world have a shared responsibility for, and a common economic interest in, protecting the global transportation network. All that remains is for these firms to begin using the Total Security Management approach to create value.

Notes

1. Leslie Woolf, "The Future of Global Trade," U.S. *Customs and Border Protection Today on the Web*, March 2005, <http://www.cbp.gov/xp/CustomsToday/2005/March/global_trade.xml> (20 April 2006).
2. Economist.com, *Research Tools Economics A-Z*, 2006, <http://www.economist.com/research/Economics/alphabetic.cfm?LETTER=F#FREE%20RIDING> (28 April 2006).
3. Deloitte Research, *Prospering in a Secure Economy*, 2004, <http://www.deloitte.com/dtt/cda/doc/content/DTT_DR_ProsSecFull_Sept2004.pdf#search='Deloitte%20Research%20prospering%20in%20the%20secure%20economy> (10 April 2006).
4. "New port security legislation introduced," *Logistics Management*, 16 March 2006, <http://www.logisticsmgmt.com/article/CA6316679.html?nid=2756> (21 April 2006).

5. Kelby Woodard, "A Strategy of Trust," *Prevention Magazine*, November-December 2004, <http://www.cargosecurity.com/ncsc/ncsc_dotnet/articals/Global%20_Supply_Chain.pdf> (13 April 2006).
6. Stanford University Global Supply Chain Management Forum, "Innovators in Supply Chain Security: Better Security Drives Business Value," The Manufacturing Institute, July 2006, <http://www.nam.org/supplychainsecurity>.

Risk and Vulnerability Assessment Report Template

The following template is a sample of a generic vulnerability assessment report. This template should be adapted for use by any specific firm in the process of conducting a vulnerability assessment.

Table of Contents

IV. Conclusions
    Observations
    Recommendations
    Additional Solutions
    Additional Issues

I. Executive Summary

_____ performed a risk assessment and threat analysis during the period of _____. Through this process, we examined factors affecting _____'s risk management strategy, including physical, processing, and information system vulnerabilities to natural and manmade disruptive events as well as several broader, industry-wide concerns that could affect normal business procedures.

Our risk assessment process is predicated on the understanding that risk is a function of two factors: the likelihood of an event taking place and the severity of such an event should it occur. Accordingly, we began this assessment by evaluating threats facing the _____ industry as a whole as well as those specific to _____, and then conducted a physical site visit to analyze the potential severity of existing vulnerabilities.

Overall, our conclusions are that _____ is/is not well positioned to handle the challenges presented by the foreseeable disruptive events that arise in the normal course of conducting business. For example, _____'s physical site is/is not appropriately protected against unauthorized vehicular access, has/does not have a thorough process for inspecting incoming deliveries, and is/is not able to continue operations for approximately _____ days based on current inventories and supplies.

Our assessment identified the following key areas for improvement and additional items for further management consideration.

Recommend _____ review and document all emergency procedures (medical, fire, disgruntled employee, and suspicious package) and train employees on emergency response instructions. These instructions should be bilingual and publicly posted, as appropriate.

Recommend _____ install closed-circuit television (CCTV) cameras and any necessary lights for monitoring and recording activity in the shipping area where trucks are loaded (the presence of these cameras will in large part deter anyone from using the company's trucks for smuggling contraband).

Recommend _____ consider upgrading its information technology (IT) access control processes, to include implementing modern biometric data for user authentication and appropriate system security for remote access (for example, traveling sales force/telecommuting) by authorized users.

In addition, we identified several additional considerations in order to ensure optimal business continuity preparedness, including:

- Consider [e.g., itemizing contractual issues to strengthen the wording in customer and vendor contracts to ensure that the firm maintains priority status for delivery during crises...]
- Consider [e.g., itemizing physical recommendations such as identification of an appropriate site for auxiliary power units to be brought in to run the plant's operations...]

II. Assessment Background

_____ performed a risk assessment and threat analysis for _____ during the period of_____. Through this process, we examined factors affecting the client's risk management strategy, including physical, organization, and value chain vulnerabilities to natural and manmade disruptive events, as well as several broader industry-wide issues and concerns that could affect normal business procedures.

Manmade and Natural Threats

Following the tragic events of September 11, 2001 Americans, and American businesses, were prompted to think anew about disruptive events and their potentially devastating effects on our normal routines. The terrorist threat continues to evolve, and it has been well documented that a mainstay of Osama Bin Laden's Al Qaeda is to strike at the American economy. Bin Laden himself declared, as recently as April 2004, "After the strike of the New York blessed days, thanks to God, their losses exceed a trillion dollars...their budgets have been in deficits for the third year in a row." He continued the point in November 2004, saying, "Even more dangerous and bitter for America is that the mujahidin recently forced Bush to resort to emergency funds to continue the fight in Afghanistan and Iraq, which is evidence of the success of the bleed-until-bankruptcy plan—with Allah's permission."

Disruptive events include more than just terrorist attacks, significant though this threat may be. They include both accidental and rolling power outages, labor disputes that create work stoppages at major seaports, and natural disasters. Significant examples of each of these events have occurred since 9/11. History also tells us naturally occurring disease outbreaks such as SARS and foot-and-mouth disease will continue to occur, affecting the free flow of goods at the nations' borders. Medical experts also warn that we are nearly a decade overdue for a global flu pandemic, the last of which occurred in 1918 and killed 675,000 Americans—more than all the wars of the last century combined.[1] The effects of such disruptions on all facets of life, including the free flow of commerce, would be immense.

Crime Concerns

[Include relevant industry and value chain partner security considerations]

Industry Overview

[Discussion of industry, seasonal sales cycles, and critical nodes in the value chain and supplier networks. Also discuss relevant U.S. and international security and regulatory requirements]

Corporate Overview

[Corporate overview, size of operation, location of facilities]

Value Chain Overview

[Discussion of significant Value Chain partners and risk assessment considerations thereof]

III. The Risk Assessment Process

Our approach to risk assessment separates terrorism and other disruptive events into two key elements: the likelihood of an industry being affected by disruptive events and the severity of such disruptive events upon the firm and its value chain partners. This approach promotes effective, predictive risk management by identifying known risks to a company, its industry, and all associated value chain partners. It also facilitates deeper analysis and focus on previously unrecognized threats emanating from the evolution of new forms of risk and a broader understanding of the systemic effects of the degradation or destruction of key critical infrastructures (power, water, transportation, etc.).

Risk = Likelihood of Attack * Severity

These key components of risk do not act in isolation—in fact, they are closely dependent on and reinforce each other. For instance, co-locating your key assets will make a single attack potentially all the more devastating (as with the New York City government's decision to co-locate most disaster recovery offices and many supplies in and around the World Trade Center complex). At the same time, reducing vulnerability to attack by hardening the facility with increased physical protection should in turn make the targeting of the facility less likely.

Perhaps the most important part of the analysis is examining the potential for cascading effects of a single event, such as how loss of power may preclude the pumping of water to fire control systems which, by law, means plant activities must cease. Other examples include the ways in which a seaport's closure can disrupt the flow of parts, preventing the assembly and distribution of final products. This analysis of the physical and logical interrelationships of all the components, both within and outside the organization, is the basis for understanding and avoiding potentially catastrophic events that, on the surface, may not seem as high risk as other events. It is these key processes that must be protected from harm in order to most efficiently return to normal operations. Identifying and mitigating cascading effects is one of the primary functions of the analysis team.

Catalog of Assets

_____ began the assessment by determining which assets are critical to _____'s business processes, evaluating threats facing the _____ industry as a whole, as well as those specific to _____, and then conducting a physical site visit to analyze the potential severity of existing physical, process, and information systems vulnerabilities.

We identified _____ critical operational concerns for _____:

· Equipment: Support functions and product processing.
· People: Key employees have process knowledge and credibility with staff.
· Information: Proprietary pricing, customer lists, personnel data, production, and marketing plans.
· Reputation: Protecting brand name and company position.

Analysis of Threats

The primary threats facing _____ are:
[Discussion of company-specific and industry-relevant issues such as closure of ports, loss of power, loss of key personnel, workplace violence]

Analysis of Vulnerabilities

_____'s primary survey areas for _____ included:
[Discussion of issues such as:

· Market risk (peer competitors, threats to industry)
· Value chain processes (inputs and outputs, redundancies in packaging and processing supplies)
· Access (gates, barriers)

- Security equipment (lights, cameras, inspection stations)
- Security procedures (visitor/access control, security force)]

IV. Conclusions

Observations

[Discussion of the industry's primary drivers for the cost of production, including such issues as worldwide supply and demand factors, ease of cross-border movement of goods, cost of production, and various regulatory or other compliance issues. Include major threats to the requisite transportation network, and the potential for cascading impact upon the value chain's process inputs and product delivery processes.]

[Include analysis of the firm's readiness relative to significant disruptive events.]

Recommendations, Additional Solutions, and Additional Issues

[Provide fuller explanation, including associated cost figures, for the recommendations outlined in the executive summary.]

Notes

1. The BBC declared on 28 November 2003, "The world will be in deep trouble if the impending influenza pandemic strikes." (Officials Reject Flu Risk Claims, <news.bbc.co.uk/1/hi/health/3244464.stm>). Two days later USA Today concurred, reporting, "The warning sirens are screaming: A deadly, contagious strain of flu will emerge, possibly soon, flu experts say, and the world is not ready to deal with it." (Next Flu Pandemic Could Wreak Global Havoc), www.usatoday.com/news/health/2003-11-30-flu_x.htm>).

Sample Value Chain Security Procedures for Chemical Firms

I. Purpose

The process outlined here will reduce the likelihood and impact of a wide range of threats including vandalism, sabotage, workplace violence, and terrorism. To assess and address potential security risks within the value chain, ensure all value chain service providers have appropriate security programs in place for minimizing the risk of a significant security incident involving the company's materials or means of transportation, as well as for maintaining consistent senior-level corporate awareness of value chain security matter.

II. Scope

All value chain operations and their associated risks, including transport and receipt of inbound materials, the transport, distribution and warehousing of outbound materials, and toll manufacturing processes of any security-sensitive materials.

III. Responsibilities

The CEO maintains overall responsibility for ensuring viability of the supply chain security.

The VP of Health and Safety (VP/HS) prepares and annually updates the list of security-sensitive materials and defines value chain partner qualification questions.

The VP of Strategic Sourcing is responsible for coordinating interaction with and participation by suppliers of security-sensitive materials.

The Director of Logistics is responsible for all carrier, service provider, and warehouse qualification procedures.

Business-Specific Operations Directors are responsible for all aspects relating to the toll manufacturing of security-sensitive materials.

Product Stewards must consider security for any security-sensitive materials as part of the Life Cycle Assessment (LCA) process.

Value Chain Managers (with assistance from VP/HS) must ensure that the requisite security parameters are included as part of their value chain partner qualification procedures.

IV. Definitions

- Security-Sensitive Materials—These are materials that, in the event of a malicious act, such as a terrorist attack, may cause a significant risk to human health or the environment. These materials include all substances covered by the Chemical Weapons Convention and those that are considered as potential weapons of mass destruction by the FBI, are classified by DOT as explosive, poison gas, flammable gas, poison by inhalation hazard, or highly reactive (i.e., pyrophoric).
- Value Chain—The value chain includes all the people and companies involved in the process of moving raw materials, unfinished goods, and finished goods through the production cycle to the manufacturing site—and ultimately to the customer. It includes material/raw material supply, transportation, toll manufacturing, distribution, warehousing, importation, and export.

- Value Chain Manager—This is a senior member of the supply chain operations, who has responsibility over one or more aspects of the value chain.
- Vulnerability—This is a weakness that can be exploited in order to gain unauthorized access or provide an opportunity to cause harm, whether in disruption of business processes, negative publicity, or even loss of life.

V. Procedures

a) Security-Sensitive Materials Identification

VP/HS prepares an updated, categorized list of security-sensitive materials annually. The VP/HS then evaluates the potential severity of the consequences of an attack or other incident, ranks these security-sensitive materials accordingly, and provides this listing to each value chain manager for review and appropriate action. This list is due to the value chain managers by January 31 of each year.[1]

b) Raw Materials and Products Identification

Each value chain manager determines, based on the information provided by VP/HS, if a security-sensitive material is being used within his or her area and then prepares a prioritized list of security-sensitive materials in their value chain. This list identifies the materials that are of significant enough concern to merit a vulnerability assessment in writing by name (common, product, raw material, or trade name) to the VP/HS. This list is due to the VP/HS by 28 February of each year.

Vulnerability Assessment

The vulnerability assessment identifies which process functions, such as transportation and logistics plans for example, place security-sensitive materials at greater risk. For the vulnerability assessment process to be fully effective, the entire value chain

---

[1] *Note:* All dates herein are notional to represent set periodicity and a fixed annual review cycle.

must be evaluated for vulnerabilities to attack and for other disruptions. This includes events that might not be specifically directed at the company or its products, but which may nonetheless affect their key shipping and transportation partners. Therefore, in most instances this will involve discussing security and protection procedures with suppliers, warehouse managers, distributors, and transportation and logistics managers.

Understanding and assessing vulnerabilities also involves determining relative values for the products and processes that rely on certain key materials. This is because if the security-sensitive material is a key component of a critical company output, then it is of higher value and therefore may merit higher protective measures. This is accomplished by categorizing relative risk for each security-sensitive material as low, medium, or high.

There are specific characteristics that make materials more or less vulnerable, and these include:

· Degree of unauthorized access to the shipments.
· Degree of supervision of the shipments.
· Visibility and location of the shipments.
· Predictability of shipments.
· Special procedures or controls already in place.
· Special equipment design (e.g., containers for aluminum alkyls).
· Whether such products/raw materials are shipped in single packaging greater than their threshold volumes.

These vulnerability assessments are to be completed by April 30th of each year.

Action Item Matrix (AIM) Development

Any security-sensitive materials that have passed through the hazard screen described previously, and for which vulnerability has been established, must be evaluated to determine whether any additional mitigation controls are appropriate. The value chain manager assembles a review team composed of representatives from the security, logistics, Environmental, Health and

Safety (EHS), engineering, and strategic sourcing departments (and/or external subject matter experts) to identify and evaluate current risk mitigation plans and determine necessary additional measures. The factors to be considered for various proposed solutions include the likely reduction in threat of a successful attack, implementation and maintenance costs, the impact on daily operations, and any identifiable negative externalities, such as decreased efficiency or increased risk to other assets. The value chain manager then identifies these solutions on the action item matrix.

Potential solutions may include:

· Selecting transportation carriers that meet higher safety protocols.
· Developing alternative supply routes for key materials.
· Improving awareness of employee and visitor locations inside the facility.
· Installing better exterior lighting.
· Constructing perimeter barriers.
· Altering the security guard rotation to allow for a larger evening staff.
· Communicating with local authorities to ensure mutual understanding of emergency response protocols.

This process should be completed by May 30th of each year.

Action Plan Implementation

Based on the previous evaluation, and in accordance with a written plan approved and endorsed by senior management, the parties responsible for any additional specific actions must ensure implementation of the risk reduction strategies.

Evaluation and Follow–up

It is the responsibility of each value chain manager to:

· Prepare an annual summary of the security actions taken.
· Maintain a file of the annual reports.

- Track performance of specified plans.
- Take corrective actions, if required.

Ensure procedures are updated as new value chain partners are identified.

Required Documentation

Documentation may include:

- Qualification documentation for qualifying new value chain partners.
- Annual security-sensitive materials list from VP/HS
- Annual status reviews by the value chain managers as defined above.
- Completed Action Item Matrix and associated action plan.
- Information on security incidents to the company, the industry or related industries and any threats that may assist in conducting the vulnerability assessment.
- Documented review of progress in implementing the agreed upon risk mitigation measures.

VI. Education and Training

Training on the content of this procedure is self-directed for persons with specific duties and responsibilities. The VP/HS, the Vice President of Strategic Sourcing, and the Director of Logistics further shall ensure that any persons delegated a portion of their assigned responsibilities read this procedure. As necessary, additional training by internal or external experts in vulnerability assessments or other areas may be provided for those in the value chain management process.

Seven Signs of Possible Terrorist Activity

The New York State Metropolitan Transportation Authority Police Department has identified seven signs of terrorist activity—any combination of which may help a firm to better assess the state of threat to its business. These pre-incident indicators may come months or even years apart, but paying attention to the clues the enemy leaves in advance of the next attack may well save both lives and dollars. If you witness any of the following activities do not attempt to interfere directly, but rather observe every detail so you can report accurately to the authorities. Every fragment of information, no matter how insignificant it might seem, can be of use to the law enforcement community in preventing future attacks.

I. Surveillance

If there is a specific target that terrorists have chosen, that target area will most likely be under observation by the terrorists during the planning phase of the operation. They do this in order to determine the strengths, weaknesses, and number of personnel that may respond to an incident. Routes to and from the target are usually established during the surveillance phase. Pay attention to people who may be recording or monitoring activities, drawing diagrams or annotating on maps, using vision-enhancing devices, or who are in possession of floor plans or blueprints of places such as high-tech firms, financial institutions, or government/military facilities.

II. Elicitation

Elicitation is the attempt to gain information about a person, place, or operation. An example is someone attempting to gain knowledge about a critical infrastructure like a power plant, water reservoir, or a maritime port. Terrorists may attempt to research bridge and tunnel usage, make unusual inquiries concerning shipments, or inquire into a military base's operations. They may also attempt to place people in jobs with access to sensitive work locations.

III. Tests of Security

Tests of security are another means that terrorists can use to attempt to gather data. These tests are usually conducted by driving by the target, moving into sensitive areas, and observing security or law enforcement response. Terrorists running these tests are interested in the time in which it takes to respond to an incident or the routes taken to a specific location. They may also try to penetrate physical security barriers or procedures in order to assess strengths and weaknesses. They often gain legitimate employment at key locations in order to monitor day-to-day activities.

IV. Acquiring Supplies

Acquiring supplies may be a case where someone is purchasing or stealing explosives, weapons, or ammunition. It could also be someone storing harmful chemicals or chemical equipment. Terrorists would also find it useful to have in their possession law enforcement equipment and identification, military uniforms and decals, as well as flight passes, badges, or even flight manuals. If they can't find the opportunity to steal these types of things, they may try to photocopy IDs, attempt to make passports or other forms of identification by counterfeiting. Possessing any of these would make it easier for one to gain entrance into secured or usually prohibited areas.

V. Suspicious People Who Don't Belong

Observing suspicious people who just "don't belong" is another important indicator. This does not mean profiling individuals, but rather the *behavior* of individuals. This may mean someone in a workplace, building, neighborhood, or business establishment that does not fit in because of their demeanor, their language usage or unusual questions they are asking.

VI. Dry Runs

Before execution of the final operation or plan, terrorists will often conduct a practice session to identify unanticipated problems. A dry run is often an immediate precursor of an attack. Examples of this activity include such activities as monitoring a police radio frequency and recording emergency response times or mapping out routes to determine the timing of traffic flow. Multiple dry runs are normally conducted at or near the target area.

VII. Deploying Assets/Getting Into Position

The seventh and final sign or signal to look for is someone deploying assets or getting into position. Because terrorists often work in groups this phase could be marked by the arrival of multiple vehicles at the target location and often includes lookouts and other surveillance activities. Anomalies may include wearing heavy coats on a hot day to conceal weapons strapped to the terrorists' bodies. This is a person's last chance to alert authorities before the terrorist act occurs.

Sample Procedures for Handling Suspicious Packages

- Report any suspicious mail or packages to security officials immediately.
- Never cut tape, strings, or other wrappings on a suspect package or immerse a suspected letter or package in water. Either action could cause an explosive device to detonate.
- Wash hands thoroughly with soap and water.
- If possible, isolate the area where the package is located.
- If object has already been moved, place letter or package in a plastic bag or some other container to prevent leakage of contents. If not certain whether the package has been moved, avoid touching or moving the suspicious package or letter.
- Make a list of personnel who were in the room or area when the suspicious envelope or package was recognized (in case they may have been exposed to harmful substances).
- Suspicious characteristics include:
  - Unusual or unknown place of origin
  - Handwritten labels, foreign handwriting, or misspelled words
  - Incorrect titles or title with no name
  - Abnormal or unusual size or shape
  - Differing return address and postmark
  - Protruding strings, aluminum foil, or wires

- Unusual odor
- Evidence of powder or other contaminants
- Crease marks, discoloration, or oily stains
- Ticking, beeping, or other sounds
- Packages marked with special instruction such as "Personal" or "Confidential"

Adapted from Chairman of the Joint Chiefs of Staff Guide 5260: Antiterrorism Personal Protection Guide: A Self-Help Guide to Antiterrorism, October 2002.

Total Security Management Glossary

access control   Systems that restrict access to a facility or computer system using authorization rules based on the positive identification of users.

benchmark   A standard used for comparison.

best practices programs   Initiatives or activities that are considered leading edge or models for others to follow, and are considered worthy of implementation in an organization. Often used as a standard that an organization is measured against.

biometric verification   A technology which uses a biological trait to identify a person.

business continuity planning   A proactive risk management methodology used to create contingency plans for ensuring the survivability of an organization following a disruption. Also referred to a Business Continuity Plan (BCP), it provides detailed steps for resuming critical business functions following a serious disruption, such as a natural disaster.

collaboration   The act of working together to improve business processes.

coordination   Fully integrated, seamless interaction among all entities that comprise the supply chain.

cooperation   Joint action taken for the mutual benefit of all parties.

consultation   The act of seeking information in order to improve business and security.

critical infrastructure   Important physical and technological assets, the lack of which pose significant disruption to the supply chain, as well as society.

customs   A federal agency authorized to monitor imports and exports.

defense in depth   The practice of layering defenses or placing multiple barriers between an organization's critical assets and a disruption to provide added protection.

delay   The act of temporarily impeding or hindering the onset of a threat to the supply chain.

detection   The ability to discover the existence of a threat in the supply chain.

deterrence   The ability to discourage or prevent a disruption to the supply chain.

dispatching   The act of responding to a supply chain threat.

emergency response plan   A plan detailing the steps to take when responding to a crisis. The plan includes notification systems, evacuation routes, accounting for employees, and training.

fixed assets   These are permanent assets of significant value and they are difficult to replace if rendered useless.

just-in-time   A strategy for minimizing product and supply inventories in which raw materials, component parts, and even finished retail goods are ordered for delivery from suppliers. They are ordered as close as possible to the actual time of need, which reduces the cost of maintaining inventories and improves the return in investment.

metrics   A standard of measurement used to determine whether an organization is effective in a certain area, such as security.

mitigation   The act of reducing or eliminating risk.

outreach   Communicating to raise awareness about the enterprise and its activities among the stakeholders in the community and throughout the value chain.

preparedness   A state of being prepared for action.

recovery   A return to normal from a crisis situation.

resilience   Ability to recover quickly following a disruption.

RFID (Radio Frequency Identification Device)   Emerging technological device that enables a firm to very cheaply track its goods through a miniature antenna affixed to the box, container, or vehicle.

risk management   The process of evaluating the risk associated with an organization.

risk matrix   A tool for conducting risk assessments and presenting the findings, showing the severity of the consequences and the probability of mishap.

situational awareness   The act of collecting and processing information for the benefit of the decision makers who, in turn, set the priorities of a firm.

supply chain security   The process used to secure against network disruptions throughout the supply chain.

threat assessment   The process of identifying potential threats, conducting a probability analysis of threats, and forecasting the impact of threats.

Total Quality Management (TQM)   The core philosophy of continuous quality improvements, based upon the work of Dr. W. Edwards Deming.

Total Security Management (TSM)   The business practice of developing and implementing comprehensive risk management and security practices for a firm's entire value chain.

value   The result of conducting a cost/benefit analysis.

value chain   The activities of an organization that add value to the organization, including all partners in the supply chain and other stakeholders that affect the firm's value.

variability   The degree of change from consistency in the supply chain.

velocity   The speed of output in the supply chain.

visibility   The ability to identify movement in the supply chain.

vulnerability   The level of exposure to disruption in the supply chain.

Index